Fractals

Selected Reprints

Edited by Alan J. Hurd

Published by the American Association of Physics Teachers

Preface

Two years ago I first compiled the list of references in "Resource Letter FR-1: Fractals"; already the field has changed greatly. Topics I judged too new to be included in the reprint book in 1987, such as multifractals, have now blossomed into large-growth industries. New phenomena continue to come to light, yet the critical experiment that started much of the furor about fractals, the simulation of diffusion-limited aggregation by Witten and Sander in 1981, remains largely unexplained.

Fractals is intended to be an introduction to fractals in the physical sciences, a convenient collection of often-cited papers, and a reference point for locating further material. The selection criteria included length limitations, automatically disqualifying some of the key papers. In fact, the chosen material is not necessarily the "best" but is intended to be representative of the large amount in a given area.

Through lively correspondence with some of the authors, additions and substitutions have been made to better fulfill the goals of the book. Reference 164 in the Resource Letter, "Density of States on Fractals: < <Fractons> >" by S. Alexander and R. Orbach, does not appear here; it has been replaced by a more readable essay, "Linear Problems on Percolation Clusters" by S. Alexander. Reference 169, "Anomalous Diffusion on Percolation Clusters" by Y. Gefen, A. Aharony and S. Alexander is included, though it is not marked by an asterisk in the Resource Letter (a glaring omission on my part). Finally, a rich and elementary introductory paper, "Fractal geometry: What is it, and what does it do?" by B.B. Mandelbrot, has been added; it is not cited in the Resource Letter.

I am deeply indebted to the staff at the American Association of Physics Teachers, especially Holly Rumpler, for handling the myriad details of putting this volume together for publication. My sincere thanks extend as well to Benoit Mandelbrot for his keen interest in this project.

Alan Hurd
Albuquerque, New Mexico
October 5, 1989

Fractals

Published by:

American Association of Physics Teachers
5112 Berwyn Road
College Park, MD 20740, U.S.A.

Cover Design:

The cover, which reflects one segment of the fractal pattern "Koch Island" (illustrated by James E. Martin), can be seen in Alan J. Hurd's Resource Letter on page 2. The design was executed by Rebecca Heller Rados.

ISBN #: 0 917853 37 7

Contents

RESOURCE LETTER

Roger H. Stuewer, *Editor*
School of Physics and Astronomy, 116 Church Street
University of Minnesota, Minneapolis, Minnesota 55455

This is one of a series of Resource Letters on different topics intended to guide college physicists, astronomers, and other scientists to some of the literature and other teaching aids that may help improve course content in specified fields. No Resource Letter is meant to be exhaustive and complete; in time there may be more than one letter on some of the main subjects of interest. Comments on these materials as well as suggestions for future topics will be welcomed. Please send such communications to Professor Roger H. Stuewer, Editor, AAPT Resource Letters, School of Physics and Astronomy, 116 Church Street SE, University of Minnesota, Minneapolis, MN 55455.

Resource Letter FR-1: Fractals

Alan J. Hurd
Chemical Physics Division 1152, Sandia National Laboratories, Albuquerque, New Mexico 87185

(Received 7 July 1988; accepted for publication 7 July 1988)

This Resource Letter provides a guide to the literature on fractals. Although "fractal" is a relatively new term in science, unifying many new ideas with established ones, its wide application and general popularity have made it one of the fastest growing fields in statistical physics. The letter E after an item indicates elementary level or material of general interest to persons becoming informed in the field; the letter I, for intermediate level, indicates material of somewhat more specialized nature; and the letter A indicates rather specialized or advanced material. An asterisk (*) indicates those articles to be included in an accompanying Reprint Book.

I. INTRODUCTION

The 1987 Wimbledon matches on television were interrupted by a slick new IBM commercial with alienlike fractal landscapes fading to earthly scenes of scientists working on oil recovery. Evidently, fractals have won over Madison Avenue. The realistic randomness of the computerized landscapes, which were the creation of Richard Voss and Benoit Mandelbrot of IBM and R. Greenberg Associates, belied the strict rules controlling their generation. The beauty of fractals is accessible not just to scientists and engineers but to everyone who has an eye for art. Hollywood has not ignored this fact, as fractal sequences in "Vol Libre," "Star Trek II," and "The Last Starfighter" attest. Figures 1 and 2 show extreme examples of fractal patterns, from perfect order to randomness.

As a serious discipline of study, the intellectual beauty of fractals has led to a small revolution in statistical physics over the last decade. "Fractional dimensions" have been studied for over 50 years, although the term "fractal" was not coined until 1975 by Mandelbrot. Large numbers of physicists became involved around 1980. Fractals are "hybrids" between familiar shapes and typically have noninteger dimensions. The mass of a "mass fractal" is proportional to its radius to a power between 1 and 3; this power is 3 for a solid ball, 2 for a plane, and 1 for a line. Similarly, the area of a "surface fractal" scales with the radius of the object with a nonintegral power. Such counterintuitive behavior is traced to the property of "self-similarity," which means that any portion of the fractal that is magnified by an arbitrary factor looks the same as the original fractal. This "dilation symmetry" holds in a literal sense for the regular "surface fractal" in Fig. 1 and in a statistical sense for the random "mass fractal" in Fig. 2. The scaling properties of fractals can be used with great effectiveness in analyzing physical problems.

While too new to have an established place in a science curriculum, there exists enough introductory and advanced material in the literature on fractals to piece together an exciting, informative "section" of a course at almost any level. The subject is particularly well suited for thermodynamics, statistical mechanics, and applied mathematics courses at the undergraduate level. For computer science students, fractals would be a welcome diversion while learning graphics, image analysis, or simulation techniques. The deeper results from the field deserve mention in all graduate courses dealing with polymer physics, percolation, aggregation, and turbulence among other topics. Even at the high-school level, fractals could be a valuable subject. Owing to their strong visual impact, a growing body of attractive books and films now exists for the high-school mathematics or science teacher who knows where to find it. This Resource Letter was written to aid these educators, as well as to help researchers who wish to come up to speed by an afternoon or two in the library. Therefore, preference was given to review papers. Although many research papers are cited, there are many more that can be uncovered in the bibliographies of the review papers.

Several of the references listed have been used by James E. Martin and myself for teaching a short course on fractals at an advanced undergraduate level at the last two Fall Meetings of the Materials Research Society.

Fig. 1. "Koch Island," created by repeated deletions and additions of smaller and smaller squares. (Illustration by James E. Martin.)

Other material has been included from a portion of a graduate materials science course that we have taught in conjunction with Douglas M. Smith at the University of New Mexico. The enthusiasm by which students digest the information is testament to the inherent interest in fractals, whose basic features are easily understood.

To follow the organization of topics below, it was necessary to cite a few references more than once. Some important papers are not cited at all because either I do not know about them or I have judged them to be too new or too specialized to have diffused through the fractals community. I apologize in advance to the authors of these omitted papers.

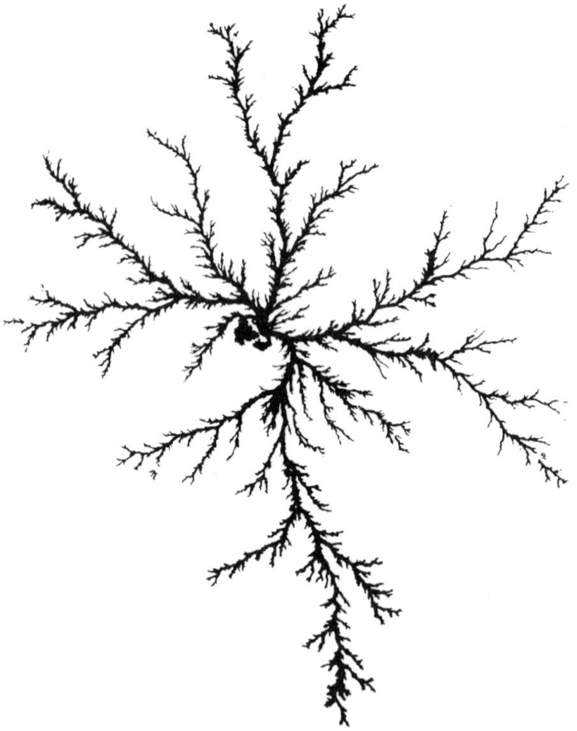

Fig. 2. Dissolution pattern in plaster. Water was injected into the center of a plaster sheet leading to the branched channels where plaster was dissolved. (Photo by G. Daccord and R. Lenormand, Ref. 96.)

II. GENERAL REFERENCES

A. Books

1. **The Fractal Geometry of Nature**, B. B. Mandelbrot (Freeman, New York, 1983). The closest thing to a true textbook on fractals, this popular book by the inventor of the subject contains a wealth of figures, which are a must in courses at all levels, as well as a discussion of a wide range of phenomena. Historical aspects and precise definitions are covered here as nowhere else, but it has no problem sets. Some sections are advanced and difficult to follow; others are elementary and very clear for all levels. Pertinent chapters are cited below. (E,I,A)
2. **The Beauty of Fractals, Images of Complex Dynamical Systems**, H.-O. Peitgen and P. H. Richter (Springer-Verlag, Berlin, 1986). Many color and black-and-white pictures highlight the discussion of complex dynamics in this book. A must for the coffee table and bookshelf alike. Algorithms for playing with fractal sets are included as well as an essay on the relation of science to art. (E,I,A)
3. **Aggregation and Fractal Aggregates**, R. Jullien and R. Botet (World Scientific, Singapore, 1987). Good technical discussion of aggregation phenomena and review of simulation and experimental results. Color plates illustrate important models. (I,A)
4. **On Growth and Form, Fractal and Non-Fractal Patterns in Physics**, edited by H. E. Stanley and N. Ostrowsky (Martinus Nijhoff, Dordrecht, The Netherlands, 1986). An outgrowth of the Cargése NATO school in Corsica, France, in 1985, this volume is both a textbook and a conference proceedings. Eleven pedagogical lectures make up "The Course" and 20 or so research papers make up "The Seminars." Several important review articles in The Course are cited below. (I,A)

B. Conference proceedings

5. **Physics of Finely Divided Matter**, edited by M. Daoud and N. Boccara (Springer-Verlag, Heidelberg, 1985). A collection of invited review and research papers; generally at a very high level. Proceedings of the Les Houches Conference. (A)
6. **Statistical Physics, Invited Lectures from STATPHYS 16**, edited by H. E. Stanley (North-Holland, Amsterdam, 1986). Although STATPHYS 16 (Boston, 1986) was not specifically a fractals conference, the large number of topics dealing with fractals in this volume demonstrates their importance to current thinking in statistical physics. Reprinted from Physica A **140** (1986). (I,A)
7. **Kinetics of Aggregation and Gelation**, edited by F. Family and D. P. Landau (North-Holland, Amsterdam, 1984). An early conference with many first-time results. (A)
8. **Fractal Aspects of Materials: Metal and Catalyst Surfaces, Powders and Aggregates**, edited by B. B. Mandelbrot and D. E. Passoja (Materials Research Society, Pittsburgh, 1984). These "extended abstracts," from the first Materials Research Society symposium on fractals, emphasize the use of fractal analysis in materials. Some of the most novel applications appear in this context. (I,A)
9. **Fractal Aspects of Materials**, edited by R. B. Laibowitz, B. B. Mandelbrot, and D. E. Passoja (Materials Research Society, Pittsburgh, 1985). Extended abstracts from the 1985 Fall Meeting of the Materials Research Society in Boston. (I,A)
10. **Fractal Aspects of Materials II**, edited by D. W. Schaefer, R. B. Laibowitz, B. B. Mandelbrot, and S. H. Liu (Materials Research Society, Pittsburgh, 1986). Extended abstracts from the 1986 Fall Meeting of the Materials Research Society. (I,A)
11. **Fractal Aspects of Materials: Disordered Systems**, edited by A. J. Hurd, D. A. Weitz, and B. B. Mandelbrot (Materials Research Society,

Pittsburgh, 1987). Extended abstracts from the 1987 Fall Meeting of the Materials Research Society. (I,A)

12. **Scaling Phenomena in Disordered Systems**, edited by R. Pynn and A. Skjeltorp (Plenum, New York, 1986), NATO ASI Series B **133**. Proceedings of the Geilo (Norway) conference, April, 1985. (I,A)
13. **Physics of Complex and Supermolecular Fluids**, edited by S. A. Safran and N. A. Clark (Wiley–Interscience, New York, 1987). Proceedings of the first Complex Fluids conference (Annandale, NJ, 1985). Section VI contains most of the fractal-related material. (A)
14. **Fractals in Physics**, edited by L. Pietronero and E. Tosatti (North-Holland, Amsterdam, 1986). Proceedings of the Sixth Trieste International Symposium on Fractals in Physics, ICTP, Trieste, Italy, 1985. A very complete overview of the current status of fractals. Several review and research articles are cited below. (I,A)
15. **Chaotic Dynamics and Fractals**, M. F. Barnsley and S. G. Denko (Academic, New York, 1986), Notes and Reports in Mathematics in Science and Engineering, Vol. 2. (A)
16. **Chaos, Fractals and Dynamics**, P. Fischer and W. R. Smith (Marcel Dekker, New York, 1985), Lecture Notes in Pure and Applied Mathematics, Vol. 98. (A)
17. **Physics and Chemistry of Porous Media II**, edited by J. R. Banavar, J. Koplik, and K. W. Winkler (American Institute of Physics, New York, 1987). AIP Conference Proceedings, Vol. 154. Many of the research papers presented at this conference at Schlumberger-Doll Research Laboratories deal with rocks. (A)
18. **Proceedings of a Symposium on Fractals in the Physical Sciences**, M. F. Shlesinger, B. B. Mandelbrot, and R. J. Rubin, J. Stat. Phys. **36**(5/6) (1984). National Bureau of Standards conference in Gaithersburg. (E,I,A)

III. GENERAL INTEREST AND REVIEW ARTICLES

*19. **"Fractal Growth,"** L. M. Sander, Sci. Am. **256**(1), 94–100 (1987). Excellent for explaining "screening" and pair correlation function in random fractals that show up in aggregation, viscous fingering, and dielectric breakdown. (E,I)

20. **"Growth: An Introduction,"** H. J. Herrmann, in Ref. 4, pp. 3–20 (1986). Part I of a two-part review (see Ref. 21). (I,A)
21. **"Form: An Introduction to Self-Similarity and Fractal Behavior,"** H. E. Stanley, in Ref. 4, pp. 21–53 (1986). Part II (see Ref. 20). An excellent review particularly for theoretical models (Pt. I) and for measurement techniques (Pt. II). (I,A)
22. **"Introduction to Fractal Geometry: Order in the Geometric Chaos of Roughness and Fragmentation,"** B. B. Mandelbrot, in Ref. 8, pp. 1–3. (E,I)
23. **"The Beauty of Fractals,"** H.-O. Peitgen, in *Proceedings of the SEAS Spring Meeting, 1987: Systems Architecture*, pp. 231–38 (SHARE Eur. Assoc., Nijmegen, 1987). In German. (E)
24. **"Dynamics on Fractal Structures,"** A. Aharony, Phys. Scr. (Sweden) **T13**, 43–4 (1986). Mainly random walks. (A)
25. **"A New Geometry—Fractal Geometry,"** IBM Nachr. (Germany) **35**, 72–3 (1985). In German. (E)
26. **"Dimensionality,"** J. D. Barrow, Philos. Trans. R. Soc. London A **310**, 337–46 (1983). Explains why most physical laws are expressed in $(3 + 1)$ dimensions and why some are not. (A)
27. **"Fractals and Their Applications in Condensed Matter Physics,"** S. H. Liu, in *Solid State Physics*, edited by H. Ehrenreich and D. Turnbull (Academic, Orlando, 1986), Vol. 39, pp. 207–73. A complete overview including measurement, aggregation, dynamics, and electrical properties. (E,I,A)
28. **"The Devil's Staircase,"** P. Bak, Phys. Today **39**(12), 39–45 (1986). (E,I)
29. **"Fractals in Physics: Squig Clusters, Diffusions, Fractal Measures, and the Unicity of Fractal Dimensionality,"** B. B. Mandelbrot, J. Stat. Phys. **34**, 895–930 (1984). (A)
30. **"Fractal Hamiltonians in Condensed Matter Physics,"** I. Z. Kostadinov, Phys. Scr. **36**, 516–21 (1987). The spin-glass random field problem and the problem of dynamics on fractals (see Fractons) are discussed. (A)
31. **"The Microgeometry and Transport Properties of Sedimentary,"** A. H. Thompson, A. J. Katz, and C. E. Krohn, Adv. Phys. **36**, 625 (1987). (I,A)
32. **"The Shapes of Random Walks,"** J. Rudnick and G. Gaspari, Science **237**, 384 (1987). (I,A)

IV. COMMENTARY

33. **"Math Technique Turns Chaos Into Order,"** A. Pollack, *New York Times*, 22 January 1985, p. C1. Comments on computer imagery and opinions from leading scientists. (E)
34. **"Fractals: Where's the Physics?"** Leo P. Kadanoff, Phys. Today **39**(2) 6–7 (1986). Raises interesting questions on where fractal analysis is going, and challenges the use of fractals in many applications. (E)
35. **"Multifractals and Fractals,"** B. B. Mandelbrot, Phys. Today **39**(9), 11 (1986). Mandelbrot answers Kadanoff, Ref. 34. (E)
36. **"Vibrational Anomalies Are Not Generally Due to Fractal Geometry: Comments on Proteins,"** J. A. Krumhansl, Phys. Rev. Lett. **56**, 2696–99 (1986). A critique of ideas concerning fractons and anomalous diffusion. See Refs. 37 and 165. (I,A)
37. **"Gentle Warning on Fractal Fashions,"** J. Maddox, Nature **322**, 303 (1986). A discussion of Krumhansl's paper, Ref. 36. See answer by S. Alexander in Ref. 165. (E)
38. **"Fractals—Fact or Fiction?"** J. J. Mecholsky, Earth Miner. Sci. **55**, 29–33 (1986). Containing both review material and commentary, this paper discusses materials science and geophysical issues. (E)
39. **"A Fractal Universe?"** D. E. Thomsen, Sci. News **129**, 217 (1986). (E)
40. **"Classical Continued Fractals,"** I. Stewart, Nature **318**, 512 (1985). Discussion of recent progress on the Seifert conjecture in mathematics. (E)
41. **"Fractal Growth Processes,"** L. M. Sander, Phys. Today **38**(1), S–19 (1985). (E)
42. **"The Man Who Reshaped Geometry,"** J. Gleick, *New York Times*, 8 December 1985, Sec. 6, p. 64 (7 pp. long). Article about B. Mandelbrot. (E)
43. **"Packing It In,"** I. Peterson, Sci. News **131**, 283–185 (1987). Image compression work of M. F. Barnsley is discussed. (E)
44. **"Models of the Sol-Gel Transition,"** G. Hentschel, Nature **315**, 719 (1985). (E)
45. **"Of Men and Ideas,"** B. B. Mandelbrot, in Ref. 1, Sec. 12, pp. 391–424. Lively reading; historical and biographical sketches. (E)

V. VIDEO AND SOFTWARE

46. **"Nothing but Zooms,"** Art Matrix, P.O. Box 880, Ithaca, NY 14851-0880; tel: (607) 277-0959. Thirty-minute film with music zooming in on the Mandelbrot set. $25.00 (E)
47. **"Fractal Fantasy,"** C. Fitch (Media Magic, 1987). Fractal images, animation, and software, available from Media Magic, P.O. Box 2069, Mill Valley, CA 94942; tel: (415) 381-4224. (E)
48. **"Aggregation,"** M. Kolb (ZEAM, Frei Universitat Berlin, 1985). A 25-min. film available on tape or film. For more information, contact M. Kolb, Laboratoire de Physique des Solides, Bat. 510, Université de Parissud, 91405 Orsay, France. See Ref. 14, pp. 267–68. (E)
49. **"Introduction to Fractals,"** L. Fogg, Micro Cornucopia (USA) No. 33, 36–40 (Dec. 1986/Jan. 1987). (E)
50. **"Fractals,"** P. R. Sorensen, BYTE (USA) **9**, 157–172 (1984). Contains a "hacker's kit" consisting of a BASIC program and documentation for making fractals. (E)
51. **"Gold DLAs,"** A. J. Hurd and D. G. Madeleine (Sandia National Laboratories, Albuquerque, 1987). Three-minute footage of the growth of metallic gold clusters is available on videocassette for the cost of materials and shipping. (E)
52. **"Exploring Mandelbrot Fractals with MMSFORTH,"** J. A. Miller and J. J. Gerow, J. Forth Appl. Res. **4**, 339–42 (1986). Discusses efficient FORTH algorithms for generating fractals. (I)
53. **"Generation and Display of Geometric Fractals in 3-D,"** V. A. Norton, Comput. Graph. **16**, 61–5 (1982). (E)
54. **"From Here to Reality (Computer Graphics Animation Systems),"** S. Beesley, Your Comput. (GB) **4**, 82–3 (1984). (E)
55. **"Random Fractal Forgeries,"** R. F. Voss, in *Physics-Like Models of*

Computation, Proc. NATO Adv. Study Inst. on Fundamental Algorithms in Computer Graphics (Ilkley, Yorkshire, UK, 1985); Physica **10D**, 81–5 (1984). (E,I)

VI. CHARACTERIZATION TECHNIQUES AND ALGORITHMS

Listed in this section are descriptions of techniques and algorithms for measuring the fractal dimension of an object, such as those in Figs. 1 and 2.

*56. **"Random Fractals: Characterization and Measurement,"** R. F. Voss, in Ref. 12, pp. 1–12. Some important concepts are discussed here that are hard to find elsewhere, such as self-affinity, lacunarity, and spectral densities. (I,A)

57. **"Form: An Introduction to Self-Similarity and Fractal Behavior,"** H. E. Stanley, see Ref. 4, pp. 21–53. (E,I)

58. **"The Electron Chalkboard,"** R. Williams, Intelligent Instrum. Comput. (USA) **5**, 140–46 (1987). Contains BASIC listings of programs that compute fractal dimensions. (E,I)

59. **"The Variation Method: A Technique to Estimate the Fractal Dimension of Surfaces,"** B. Dubuc, C. Roques-Carmes, C. Tricot, and S. W. Zucker, in Ref. 11, pp. 83–5. See also Ref. 10, pp. 115–17. The Variation Method is a very stable way to characterize fractal profiles. (A)

60. **"Fractal Aggregates and Their Fractal Measures,"** P. Meakin, in *Phase Transitions and Critical Phenomena*, edited by C. Domb and J. L. Laibowitz (Academic, New York, 1988), Adv. Colloid Interface Sci. **12**, 1–32. (I,A)

61. **Practical Stereology**, J. C. Russ (Plenum, New York, 1986). Chapter 7 discusses useful techniques for digitized images of fractal curves and surfaces. (E)

62. **"On the Measurement of the Fractal Dimension of Aggregated Particles by Electron Microscopy: Experimental Method, Corrections and Comparison with Numerical Models,"** M. Tence, J. P. Chevalier, and R. Jullien, J. Phys. (Paris) **47**, 1989–98 (1986). (I)

A. Multifractal analysis

63. **"Scaling Structure of the Surface Layer of Diffusion-Limited Aggregates,"** T. C. Halsey, P. Meakin, and I. Procaccia, Phys. Rev. Lett. **56**, 854–57 (1986). This and the following paper first applied the idea of multifractal analysis, originally developed in a turbulence context (Ref. 172), to diffusion-limited aggregates. (A)

64. **"Growth Probability Distribution in Kinetic Aggregation Processes,"** C. Amitrano, A. Coniglio, and F. di Liberto, Phys. Rev. Lett. **57**, 1016–119 (1986). (A)

65. **"The Infinite Number of Generalized Dimensions of Fractals and Strange Attractors,"** H. G. E. Hentschel and I. Procaccia, Physica **8D**, 435–44 (1983). (A)

See also Refs. 35 and 172.

VII. AGGREGATION

The first breakthroughs in using fractals in physics were around 1980. One was the realization by Forrest and Witten in 1979 that metal smoke aggregates might be a kind of fractal. The subsequent simulation models for aggregation quickly established that random walks are important to the geometry of many kinds of clusters. This idea united many fields together.

A. Reviews

66. **"Tenuous Structures from Disorderly Growth Processes,"** T. A. Witten and M. E. Cates, Science **232**, 1607–12 (1986). Up-to-date review of experimental, simulation, and theoretical results on aggregation and related growth processes. (I,A)

67. **"Fractal Growth,"** L. M. Sander, Sci. Am. **256**(1), 94–100 (1987). See Ref. 19. (E,I)

68. **"Computer Simulation of Growth and Aggregation Processes,"** P. Meakin, in Ref. 4, pp. 111–35. (I,A)

69. **"Geometrical Cluster Growth Models and Kinetic Gelation,"** H. J. Herrmann, Phys. Rep. **136**, 153–227 (1986). (I,A)

70. **"Superpolymers, Ultraweak Solids and Aggregates,"** T. C. Lubensky and P. A. Pincus, Phys. Today **37**(10), 44–50 (1984). (E,I)

71. **"Fractal Growth Processes,"** L. M. Sander, Nature **322**, 789–93 (1986). (E,I)

72. **"Fractal Aggregates,"** R. Jullien, Comments Condens. Matter Phys. (GB) **13**, 177–205 (1987). (I,A)

73. **"Kinetics of Clustering in Irreversible Aggregation,"** M. H. Ernst, in Ref. 14, pp. 289–302. A complete summary of what is known about aggregation via the Smoluchowski coagulation equation. (A)

74. **"Fractal Clusters,"** B. M. Smirnov, Sov. Phys. Usp. **29**, 481–505 (1986). An interesting viewpoint on aggregation, electrodeposition, and ball lightning. (E,A)

B. Canonical models, simulations, and experiments

*75. **"Diffusion-Limited Aggregation, a Kinetic Critical Phenomenon,"** T. A. Witten and L. M Sander, Phys. Rev. Lett. **47**, 1400–03 (1981). Seminal paper on discrete simulations of particle–cluster aggregation by the Witten–Sander algorithm for diffusion-limited aggregation (DLA). (A)

76. **"Diffusion-Limited Aggregation,"** T. A. Witten and L. M. Sander, Phys. Rev. B **27**, 5686–97 (1983). An expansion of Ref. 75. (A)

*77. **"Scaling of Kinetically Grown Clusters,"** M. Kolb, R. Botet, and R. Jullien, Phys. Rev. Lett. **51**, 1123-26 (1983). This seminal paper and the following one were published side by side. They defined a simulation model of cluster–cluster aggregation, which established that fractal structure should describe colloidal aggregates. (A)

*78. **"Formation of Fractal Clusters and Networks by Irreversible Diffusion-Limited Aggregation,"** P. Meakin, Phys. Rev. Lett. **51**, 1119–22 (1983). See Ref. 77. (A)

79. **"Fractal Geometry of Colloidal Aggregates,"** D. W. Schaefer, J. E. Martin, P. Wiltzius, and D. S. Cannell, Phys. Rev. Lett. **52**, 2371–74 (1984). Early scattering study using both x rays and light first demonstrated the power of fractal structure determination by scattering. (A)

80. **"Fractal Structures Formed by Kinetic Aggregation of Aqueous Gold Colloids,"** D. A. Weitz and M. Oliveria, Phys. Rev. Lett. **52**, 1433–36 (1984). Early imaging study on gold aggregates. (A)

*81. **"The Structure of Aggregated Gold Colloids,"** P. Dimon, S. K. Sinha, D. A. Weitz, C. R. Safinya, G. S. Smith, W. A. Varady, and H. M. Lindsay, Phys. Rev. Lett. **57**, 595 (1986). Gold aggregation scattering data were analyzed by a detailed scattering theory. (A)

82. **"Dynamic Scaling of Cluster-Mass Distributions in Kinetic Colloid Aggregation,"** D. A. Weitz and M. Y. Lin, Phys. Rev. Lett. **57**, 2037–40 (1986). (A)

83. **"Restructuring of Colloidal Silica Aggregates,"** C. Aubert and D. S. Cannell, Phys. Rev. Lett. **56**, 738–41 (1986). (A)

84. **"Growth: An Introduction,"** H. J. Herrmann, in Ref. 8, pp. 3–20. (I,A)

85. **"Fractal Structure of Three-Dimensional Witten–Sander Clusters,"** P. Meakin and Z. R. Wasserman, Chem. Phys. **91**, 391-97 (1984). (I,A)

86. **"Some Universality Properties Associated with the Cluster–Cluster Aggregation Model,"** P. Meakin and Z. R. Wasserman, Phys. Lett. A **103**, 337–41 (1984). (I,A)

*87. **"Diffusion-Limited Aggregation in Two Dimensions,"** A. J. Hurd and D. W. Schaefer, Phys. Rev. Lett. **54**, 1043–46 (1985). Imaging study of Brownian particles aggregating in two dimensions. (A)

88. **"Slow Aggregation of Silica,"** J. E. Martin, Phys. Rev. A **36**, 3415–26 (1987). A complete light-scattering study of reaction-limited aggregation of silica, including growth kinetics. (A)

89. **"Colloidal Aggregation Revisited: New Insights based on Fractal Structure and Surface-Enhanced Raman Scattering,"** D. A. Weitz, M. Y. Lin, and C. J. Sandroff, Surf. Sci. **158**, 147–64 (1985). Contains some review material and discussion of colloidal gold. (A)

*90. **"Fractal Dimension of Dielectric Breakdown,"** L. Niemeyer, L. Pietronero, and H. J. Wiesmann, Phys. Rev. Lett. **52**, 1033–36 (1984). The electrostatic analogy for diffusion-limited aggregation was clarified here. (A)

*91. **"Probability Scaling for Diffusion-Limited Aggregation,"** L. A.

Turkevich and H. Scher, Phys. Rev. Lett. **55**, 1026–29 (1985). See follow-up paper, Phys. Rev. A **33**, 786 (1986). Theory for DLA that exploits the electrostatic analogy. (A)

92. **"Some Recent Advances in the Simulation of Diffusion-Limited Aggregation and Related Processes,"** P. Meakin, in Ref. 14, pp. 205–12. Simulation test of the Turkevich–Scher theory, Ref. 91. (A)

C. Viscous fingers

*93. **"Fractal Growth of Viscous Fingers; Quantitative Characterization of Fluid Instability Phenomenon,"** J. Nittmann, G. Daccord, and H. E. Stanley, Nature **314**, 141–44 (1985). Good explanation of the physics in viscous fingering, with nice pictures. (I,A)

94. **"Viscous Fingering in Hele-Shaw Cells,"** P. G. Saffman, J. Fluid Mech. **173**, 73–94 (1986). Review of classical and recent results. (I,A)

95. **"Fractal Viscous Fingering: Experiments and Models,"** J. Nittmann, in Ref. 6, pp. 124–33. Review of recent results. (A)

96. **"Fractal Patterns from Chemical Dissolution,"** G. Daccord and R. Lenormand, Nature **325**, 41–43 (1987). (I)

D. Dendrite and pattern formation

*97. **"Pattern Formation in Materials Science,"** J. P. Gollub and L. M. Sander, Mater. Res. Soc. Bull. **12**(6), 98–100 (1987). The connection between crystalline dendritic growth and disorderly fractal growth is discussed. (I)

*98. **"Fractal Growth of Copper Electrodeposits,"** R. M. Brady and R. C. Ball, Nature **309**, 225–29 (1984). Important paper connecting the physics of electrodeposition to diffusion-limited aggregation. (A)

99. **"Fractal Structures of Zinc Metal Leaves Grown by Electrodeposition,"** M. Matsushita, M. Sano, Y. Hayakawa, H. Honjo, and Y. Sawada, Phys. Rev. Lett. **53**, 286–89 (1984). Imaging study on a very simple fractal system that could be reproduced in a high-school laboratory. (I,A)

100. **"Dendritic and Fractal Patterns in Electrolyte Metal Deposits,"** Y. Sawada, A. Dougherty, and J. P. Gollub, Phys. Rev. Lett. **56**, 1260 (1986). (A)

101. **"Morphology and Microstructure in Electrochemical Deposition of Zinc,"** D. Grier, E. Ben-Jacob, R. Clarke, and L. M. Sander, Phys. Rev. Lett. **56**, 1264 (1986). (A)

102. **"Instabilities and Pattern Formation in Crystal Growth,"** J. S. Langer, Rev. Mod. Phys. **52**, 1–28 (1980). Although somewhat dated, this review remains an important resource for nonfractal growth, marginal stability, and the Mullins–Sekerka instability. (I,A)

103. **"Recent Developments in the Theory of Pattern Formation,"** J. S. Langer, in Ref. 6, pp. 44–50. A complementary follow-up to Ref. 102. (A)

104. **"Random Walks with Memory,"** L. Peliti and L. Pietronero, Riv. Nuovo Cimento **10**, 1–33 (1987). (A)

105. **"Pattern Formation in Diffusion-Limited Aggregation,"** T. Vicsek, Phys. Rev. Lett. **53**, 2281–84 (1984). See comments **"No Pattern Yet for Snowflakes,"** J. Maddox, Nature **313**, 93 (1985). (E,A)

VIII. GELATION AND PERCOLATION

*106. **"Cluster Shapes at the Percolation Threshold: An Effective Cluster Dimensionality and its Connection with Critical-Point Exponents,"** H. E. Stanley, J. Phys. A: Math. Gen. **10**, L211–20 (1977). The infinite cluster in percolation is shown to be a fractal for the first time. (A)

107. **"Thermally Driven Phase Transitions Near the Percolation Threshold in Two Dimensions,"** H. E. Stanley, R. J. Birgeneau, P. J. Reynolds, and J. F. Nicoll, J. Phys. C **9**, L553–60 (1976). Probably the first physics paper to mention fractals. (A)

108. **Introduction to Percolation Theory**, D. Stauffer (Taylor and Francis, London, 1985). A complete reference book on percolation, including simulation techniques, written in a lively, sometimes hilarious style. For a shorter, older version, see Phys. Rep. **54**, 1–74 (1979). (A)

109. **"Dimensionalities and Other Geometric Critical Exponents in Percolation Theory,"** I. M. Sokolov, Usp. Fiz. Nauk (USSR) **150**, 221–55 (1986). Translated in Sov. Phys.-Usp. **29**, 924–945 (1986). (A)

110. **"Fractal Concepts in Aggregation and Gelation,"** H. E. Stanley, in Ref. 7, pp. 1–4. A summary of exponents. (A)

111. **The Physics of Amorphous Solids**, R. Zallen (Wiley–Interscience, New York, 1983). This general reference for condensed matter contains a good discussion of fractal aspects of percolation and polymers. (I,A)

112. **"Percolation and Fractals,"** A. Aharony, J. Phys. Soc. Jpn. Suppl. **52**, 125–29 (1983). (A)

113. **"Self-Similarity in Irreversible Kinetic Gelation,"** F. Family, Phys. Rev. Lett. **51**, 2112–15 (1983). (A)

114. **"Dynamic Scaling near the Percolation Threshold in Thin Au Films,"** R. Laibowitz and Y. Gefen, Phys. Rev. Lett. **53**, 380–83 (1984). (A)

IX. MATERIALS

A. Polymers

115. **"Fractal Dimensionality of Polymer Chains,"** S. Havlin and D. Ben-Avraham, J. Phys. A: Math. Gen. **15**, L311–16 (1982). An early theoretical discussion of the fractal structure of polymers. (A)

116. **"Structure of Soluble Silicates,"** D. W. Schaefer and K. D. Keefer, in *Better Ceramics Through Chemistry*, edited by C. J. Brinker, D. E. Clark, and D. R. Ulrich, Mater. Res. Soc. Symp. Proc. **32**, 1–14 (1984). One of the first applications of fractals to polymers and materials science, this work contains a good qualitative discussion of scattering techniques. (E,I)

117. **"Application of Fractal Concepts to Polymer Statistics and to Anomalous Transport in Randomly Porous Media,"** H. E. Stanley, in Ref. 18, pp. 843–60. (I,A)

118. **"Branched Polymers,"** M. Daoud, in Ref. 4, pp. 227–30. (A)

119. **Scaling Concepts in Polymer Physics**, P.-G. de Gennes (Cornell, London, 1979). A complete book on polymer physics, the text is not written in fractal language, although fractal concepts are used in the physics. (A)

B. Silica: Aerogels, vycor, and sandstone

120. **"Origin of Porosity in Synthetic Materials,"** D. W. Schaefer, J. P. Wilcoxon, K. D. Keefer, B. C. Bunker, R. K. Pearson, I. M. Thomas, and D. E. Miller, in Ref. 17, pp. 63–80. (I)

121. **"Structure of Random Porous Materials: Silica Aerogel,"** D. W. Schaefer and K. D. Keefer, Phys. Rev. Lett. **56**, 2199–202 (1986). (A)

122. **"Structure of Random Materials,"** D. W. Schaefer, J. E. Martin, A. J. Hurd, and K. D. Keefer, in Ref. 5. pp. 31–37. (I)

123. **"Surface Roughening and the Fractal Nature of Rocks,"** P.-Z. Wong, J. Howard, and J.-S. Lin, Phys. Rev. Lett. **57**, 637–40 (1986). Uses a self-affine interpretation of scattering. (A)

124. **"Fractal Sandstone Pores: Implications for Conductivity and Pore Formation,"** A. J. Katz and A. H. Thompson, Phys. Rev. Lett. **54**, 1325–28 (1985). See also Ref. 31. (A)

125. **"Fractal Geometry of Vapor-Phase Aggregates,"** J. E. Martin, D. W. Schaefer, and A. J. Hurd, Phys. Rev. A **33**, 3540–43 (1986). (A)

126. **"Surface and Mass Fractals in Vapor-Phase Aggregates,"** A. J. Hurd, D. W. Schaefer, and J. E. Martin, Phys. Rev. A **35**, 2361–64 (1987). (A)

127. **"In Situ Study of Fractal Aggregates in Fumed Silica,"** A. J. Hurd and W. L. Flower, J. Colloid Interface Sci. **122**, 178–92 (1988). (A) See also Ref. 31.

C. Biology

128. **"Physiology in Fractal Dimensions,"** B. J. West and A. L. Goldberger, Am. Sci. **75**, 354–65 (1987). (E)

129. **"Statistical Mechanics of Cellular Automata,"** S. Wolfram, Rev. Mod. Phys. **55**, 601–44 (1983). "Game of Life"-type algorithms, which are used to investigate self-organization, show self-similar patterns. (I,A)

130. **"Development, Growth, and Form in Living Systems,"** J.-P. Boon and A. Noullez, in Ref. 4, pp. 174–83. (A)

131. **On Size and Life**, T. A. McMahon and J. T. Bonner (Freeman, New York, 1983). (E)

D. Ceramics

132. **"Fractal Aspects of Ceramic Synthesis,"** D. W. Schaefer and K. D. Keefer, in *Better Ceramics Through Chemistry*, edited by C. J. Brinker, D. E. Clark, and D. R. Ulrich, Mater. Res. Soc. Proc. **73**, 277–88 (1986). Review of scattering techniques applied to structure determination. (I,A)
133. **"Growth of Fractally Rough Colloids,"** K. D. Keefer and D. W. Schaefer, Phys. Rev. Lett. **56**, 2376–79 (1986). (A)
134. **"Growth and Fractal Structure of Ceramic Precursors,"** K. D. Keefer, Mater. Res. Soc. Bull. **12**(7), 29–33 (1987). This review focuses on silicate precursors. (I)
135. **"A Systematic Description of Microstructure,"** E. Horbogen, J. Mater. Sci. **21**, 3737–47 (1986). Review. (I,A)

X. FRACTAL SURFACES

A. Scattering: Lignite, silica

*136. **"Small-Angle x-Ray-Scattering Investigation of Submicroscopic Porosity with Fractal Properties,"** H. D. Bale and P. W. Schmidt, Phys. Rev. Lett. **53**, 596–99 (1984). Seminal paper on the scattering of fractal surfaces. (A)

137. **"Fractal Surfaces in Porous Media,"** P.-Z. Wong, in Ref. 17, pp. 304–18. Review of scattering and some new things on the electrical response of rough surfaces. (I)

B. Fracture

*138. **"Fractal Character of Fracture Surfaces of Metals,"** B. B. Mandelbrot, D. E. Passoja, and A. J. Paullay, Nature **308**, 721–22 (1984). See also Ref. 144. (I)

139. **"Formation of Fractal Cracks in a Kinetic Fracture Model,"** Y. Termonia and P. Meakin, Nature **320**, 429–31 (1986). (A)

140. **"Physics of Fracturing and Seismic Energy Release: A Review,"** S. J. Gibowicz, Pure Appl. Geophys. **124**, 611–58 (1986). (I)

C. Adsorption

*141. **"Molecular Fractal Surface,"** D. Avnir, D. Farin, and P. Pfeifer, Nature **308**, 261–63 (1984). A broad survey of powdered materials. (A)

142. **"Fractal Dimension as Working Tool for Surface Roughness Problems,"** P. Pfeifer, Appl. Surf. Sci. **18**, 146–64 (1984). (I)

143. **"Applications and Limitations of Boundary-Line Fractal Analysis of Irregular Surfaces: Proteins, Aggregates, and Porous Materials,"** D. Farin, S. Peleg, D. Yavin, and D. Avnir, Langmuir **1**, 399–407 (1985). (E,I)

D. Self-affinity

144. **"Self-Affine Fractal Sets,"** B. B. Mandelbrot, in Ref. 14, pp. 3–28. A series of three lectures at the Trieste conference. (I)

145. **"Self-Affine Fractals and Fractal Dimension,"** B. B. Mandelbrot, Phys. Scr. **32**, 257–60 (1985). Clarifies different measurements applied to curves and surfaces. (I)

(See also Ref. 123.)

E. Electrical behavior

146. **"Introduction to Transfer and Motion in Fractal Media: The Geometry of Kinetics,"** A. Le Mehaute and G. Crepy, Solid State Ionics **9**, **10**, 17–30 (1983). Seminal paper that implicated fractal surfaces in the complex response of rough, nonblocking electrodes; invoked a controversial view of diffusion using a fractional derivative. (A)

147. **"Fractal Model for the ac Response of a Rough Interface,"** S. H. Liu, Phys. Rev. Lett. **44**, 529–31 (1985). A regular fractal electrode of the blocking type was studied. (A)

148. **"Fractal Electrodes and Constant Phase Angle Response: Exact Examples and Counter Examples,"** B. Sapoval, Solid State Ionics **23**, 253–59 (1987). (A)

149. **"DC and ac Response of a Nonblocking Fractal Electrode,"** B. Sapoval, in Ref. 11, pp. 66–8. (I)

XI. BEHAVIOR

A. Optical properties: Scattering

150. **"Scattering from Fractals,"** J. E. Martin and A. J. Hurd, J. Appl. Crystallogr. **20**, 61–78 (1987). This review of scattering interpretations and fractal materials includes introductory material, polydispersity, polymers, and aggregates. (I)

151. **"Neutron Scattering from Fractals,"** S. K. Sinha, Bull. Mater. Sci. **7**, 341–52 (1985). This review, which has a particularly nice introduction, touches on crossover functions and scattering from silica. (I)

152. **"Experimental Methods for Studying Fractal Aggregates,"** J. Teixeira, in Ref. 4, pp. 145–62. Critical review of scattering techniques and data fitting. (I)

153. **"Neutron Scattering from Fractals,"** J. K. Kjems, T. Freltoft, D. Richter, and S. K. Sinha, Physica B **136**, 285–90 (1986). Short review with data on the dynamics of fumed silica. (I)

154. **"Scattering Exponents for Polydisperse Surface and Mass Fractals,"** J. E. Martin, J. Appl. Crystallogr. **19**, 25–27 (1986). (A)

155. **"Properties of Fractal Colloidal Aggregates,"** H. M. Lindsay, M. Y. Lin, D. A. Weitz, P. Sheng, Z. Chen, R. Klein, and P. Meakin, Faraday Discus. Chem. Soc. **83**, 153–65 (1987). Some review, plus discussions of the optics of colloidal gold aggregates. (I,A)

156. **"Optical Diffraction on Fractals,"** C. Allain and M. Cloitre, Phys. Rev. B **33**, 3566–69 (1986). These measurements on optical Fourier transforms suggest a nice undergraduate lab experiment. (E,I)

*157. **"Optics of Fractal Clusters such as Smoke,"** M. V. Berry and I. C. Percival, Opt. Acta **33**, 577–91 (1986). A theory for scattering from clusters made of absorbing material. (A)

B. Chemical reaction kinetics

*158. **"Fractal to Classical Crossover of Chemical Reactions,"** J. S. Newhouse, P. Argyrakis, and R. Kopelman, Chem. Phys. Lett. **107**, 48–52 (1984). (A)

159. **"Reaction Kinetics of Fractals: Random-Walker Simulations and Exciton Experiments,"** R. Kopelman, W. Klymko, J. S. Newhouse, and L. W. Anacker, Phys. Rev. B **29**, 3747–48 (1984). (A)

See also Refs. 79, 81, 88, 123, and 127.

C. Geophysical

160. **"Physics of Fracturing and Seismic Energy Release: A Review,"** S. J. Gibowicz, Pure Appl. Geophys. **124**, 611–58 (1986). (I)

161. **"Area-Perimeter Relation for Rain and Cloud Areas,"** S. Lovejoy, Science **216**, 185–87 (1982).

162. **"Fractal Characterization of Inhomogeneous Geophysical Measuring Networks,"** S. Lovejoy, D. Schertzer, and P. Ladoy, Nature **319**, 43–44 (1986). (I)

163. **"Fractal Shape of Hail Clouds,"** F. S. Rys and A. Waldvogel, Phys. Rev. Lett. **56**, 784–87 (1986). (I)

D. Fractons

The vibrational aspects of fractals, or "fractons," has become a major subfield of investigation. First introduced as a theoretical concept in 1982, the work has only recently begun to explain data successfully; however, its implications are quite profound for material properties.

*164. **"Density of States on Fractals: ⟨⟨Fractons⟩⟩"** S. Alexander and R. Orbach, J. Phys. (Paris) Lett. **43**, L625–31 (1982). Seminal paper on fractons that contains, among other things, the famous Alexander–Orbach conjecture. (A)

165. **"Fractons,"** S. Alexander, in Ref. 6, pp. 397–404. This advanced theoretical review is largely an answer to comments in Refs. 36 and 37. (I,A)

*166. **"Dynamics of Fractal Networks,"** R. Orbach, Science **231**, 814–19 (1986). Wide review of the properties associated with fractons and a current critique of experimental work. (E,I,A)

167. **"Is the Elastic Energy of Amorphous Materials Rotationally Invariant?"** S. Alexander, J. Phys. (Paris) **45**, 1939–45 (1984). Discusses scalar elastic energy, including the elastic properties of the Sierpinski gasket. (A)

168. **"La Percolation: Un Concept Unificateur,"** P.-G. de Gennes, Recherche **7**, 919–27 (1976). Defines the famous "Ant-in-a-Labyrinth Problem" for diffusion on a random network. In French. (A)

169. **"Anomalous Diffusion on Percolation Clusters,"** Y. Gefen, A. Aharony, and S. Alexander, Phys. Rev. Lett. **50**, 77–80 (1983). Solves the Ant-in-a-Labyrinth Problem for fractals. (A)

170. **"Random Walks on Fractal Structures and Percolation Clusters,"** R. Rammal and G. Toulouse, J. Phys. (Paris) **44**, L13–22 (1983). Determines properties of anomalous diffusion. (A)

171. **"Brillouin Scattering Measurement of Phonon–Fracton Crossover in Silica Aerogels,"** E. Courtens, J. Pelous, J. Phalippou, R. Vacher, and T. Woignier, Phys. Rev. Lett. **58**, 128–31 (1987). An example of experimental work to observe fractons. (A)

E. Turbulence and chaos

172. **"On the Geometry of Homogeneous Turbulence, with Stress on the Fractal Dimension of the Iso-Surfaces of Scalars,"** B. B. Mandelbrot, J. Fluid Mech. **72**, 401–16 (1975). (A)

173. **"Intermittent Turbulence on Self-Similar Cascades: Divergence of High Moments and Dimension of the Carrier,"** B. B. Mandelbrot, J. Fluid Mech. **62**, 331–58 (1974). This paper and Ref. 172 are two of the most cited in the field of fractal turbulence. (A)

174. **Chaotic Dynamics and Fractals**, M. F. Barnsley and S. G. Denko (Academic, New York, 1986), Notes and Reports in Mathematics in Science and Engineering, Vol. 2. (I,A)

175. **Chaos, Fractals, and Dynamics**, P. Fischer and W. R. Smith (Marcel Dekker, New York, 1985), Lecture Notes in Pure and Applied Mathematics, Vol. 98. (I,A)

176. **"Certain Problems in the Theory of Developed Hydrodynamical Turbulence,"** E. Levich, Phys. Rep. **151**, 129–238 (1987). Major review of turbulence. (E,I,A)

177. **"Characterization of Hydrodynamic Strange Attractors,"** H. L. Swinney and J. P. Gollub, Physica D **18**, 448–554 (1986). Review of measurements in weakly turbulent flows. (I)

178. **"Chaotic Dynamics of Instabilities in Solids,"** C. D. Jeffries, Phys. Scr. **T9**, 11-26 (1985). (I)

179. **"The Dimension of Chaotic Attractors,"** J. D. Farmer, E. Ott, and J. A. Yorke, Physica D **7**, 153–180 (1983). (A)

180. **"New Global Fractal Formalism Describes Paths to Turbulence,"** B. G. Levi, Phys. Today **39**, 17–8 (1986). (I)

ACKNOWLEDGMENTS

I am grateful to many individuals for contributing to the material here, including Dale Schaefer, Dave Weitz, Richard Voss, Jim Martin, and, especially, the authors who have allowed their papers to be placed in the Reprint Book. A special thanks is in order for Dr. and Mrs. Mandelbrot for providing a large computer file with recent references on fractals.

Reprinted with permission from the *Proceedings of the Royal Society of London,* B.B. Mandelbrot, "Fractal Geometry: What is it, and what does it do?" A423, 3–16, ©1989 The Royal Society, London.

Fractal geometry: What is it, and what does it do?

B.B. Mandelbrot
Physics Department, IBM T.J. Watson Research Center, Yorktown Heights NY 10598
Mathematics Department, Yale University, New Haven CT 06520

Fractal geometry is a workable geometric middle ground between the excessive geometric order of Euclid and the geometric chaos of general mathematics. It is based on a form of symmetry that had previously been underutilized, namely invariance under contraction or dilation. Fractal geometry is conveniently viewed as a language that has proven its value by its uses. Its uses in art and pure mathematics, being without "practical" application, they can be said to be poetic. Its uses in various areas of the study of materials and of other areas of engineering are examples of practical prose. Its uses in physical theory, especially in conjunction with the basic equation of mathematical physics, combine poetry and high prose. Several of the problems that fractal geometry tackles involve old mysteries, some of them already known to primitive man, others mentioned in the Bible, and others familiar to every landscape artist.

Fractals provide a workable new middle ground, between the excessive geometric order of euclid, and the geometric chaos of roughness and fragmentation

Instead of attempting to introduce and link together the papers that follow in this Discussion Meeting, we prefer to ponder the question, "What is fractal geometry?" We write primarily for the comparative novice, but have tried to include tidbits for the already informed reader.

Before we tackle what a fractal *is*, let us ponder what a fractal *is not*. Take a geometric shape and examine it in increasing detail. That is, take smaller and smaller portions near a point *P*, and allow every one to be dilated, that is, enlarged to some prescribed overall size. If our shape belongs to standard geometry, it is well known that the enlargements become increasingly smooth. Ultimately, nearly every connected shape is locally linear. One can say for example that, "a generic curve is attracted under dilations" to a straight line (thus defining the tangent at the point *P*). And "a generic surface is attracted by dilation" to a plane (thus defining the tangent plane at the point *P*). More generally, one can say that nearly every standard shape's local structure converges under dilation to one of the small number of "universal attractors." The term "attractor" is borrowed from dynamics and probability theory, and the even more grandiose term, "universal," is borrowed from recent physics. An example of exception to this rule is when *P* is a double point of a curve; the curve near *P* is then attracted to two intersecting lines and has two tangents; but double points are few and far between in standard curves.

Standard geometry and calculus (which is intimately related to it) have long proven to be extraordinarily effective in the sciences. Yet there is no question that Nature *fails to be* locally linear. In fact, the shapes of Nature are so varied as to deserve being called "geometrically chaotic," unless proven otherwise. Unfortunately, "complete" chaos could not conceivably lead to a science. This is perhaps why many of the oldest concerns of Man, such as concerns with the shapes of mountains, clouds, and trees, or with the floods of the Nile, had not led to sciences comparable in effectiveness to the physics of smooth phenomena.

Though the term "chaos" was not used at the time, one can say that a second kind of chaos became known during the half century 1975-1925, when mathematicians who were fleeing from concerns with Nature took cognizance of the fact that a geometric shape's roughness may conceivably *fail* to vanish as the examination becomes more searching. It may conceivably vary endlessly, up and down. The hold of standard geometry was so powerful, however, that the shapes constructed so that they do not reduce locally to straight lines were labeled "monsters" and "pathological."

Between the extremes of linear geometric order and of geometric chaos ruled by "pathologies," can there be a middle ground of "organized" or "orderly" geometric chaos? The author has conceived and outlined such a ground, and gave to its study the name of "fractal geometry," the fuller name being "fractal geometry of nature and chaos." It will be argued momentarily that fractal geometry is best viewed as a geometric language, new as of 1975, which incorporates as "characters" several of the mathematical monsters of 19875-1925, and whose uses have now become so diverse, that it is possible to sort them out as poetry, strictly utilitarian prose, and high prose.

Fractals are characterized by so-called "symmetries", which are invariances under dilations and/or contractions

Broadly speaking, mathematical and natural fractals are shapes whose roughness and fragmentation *neither* tend to vanish, *nor* fluctuate up and down, but remain *essentially unchanged* as one zooms in continually and examination is refined. Hence, the structure of every piece holds the key to the whole structure.

An alternative term is "self-similar," which has two meanings. One can understand "similar" as a loose everyday synonym of "analogous." But there is also the strict textbook sense of "contracting similarity." It expresses that each part is a linear geometric reduction of the whole, with the same reduction ratios in all directions. Figure 1 illustrates a standard strictly self-similar fractal, called Sierpiński gasket. In the early days, the resulting terminological ambiguity was acceptable to physicists, because early detailed studies did indeed concentrate on strictly self-similar shapes. However, more recent developments have extended, in particular, to include self-affine shapes, in which the reductions are still linear but the reduction ratios in different directions are different. For example, a relief is nearly self-affine, in the sense that, in order to go from a large piece to a small piece, one must contract the horizontal and vertical coordinates in different ratios.

Can the process of taking parts be inverted, by replacing zooming in by zooming away from the object of interest? Indeed, many fractals, including the strictly self-similar ones, can be extrapolated to become unbounded, and they can remain unchanged when the above process is reversed and the shape is examined at increasingly rough scales. To gain an idea of the appearance of an extrapolated Sierpiński gasket, it suffices to focus on a very small piece of Figure 1, and then to imagine that this piece has been blown up to letter size, and therefore the whole gasket has become so large that its edges are beyond visibility. This illustration shows that each choice of a small piece to focus upon yields a different extrapolate.

When the Sierpiński gasket is constructed as in Figure 1, that is, by deleting middle triangles, one sees it has three properties of contracting self-similarity, as pointed out in the caption of Figure 1. These properties appear, so to speak, as "static" and "after-the-fact," but this is a completely misleading impression. Its prevalence and its being viewed as a flaw came to us as a surprise. Therefore, it is good to stop and show how, knowing the same symmetries, it is easy to reconstruct the gasket "dynamically" by a stochastic interpretation of a scheme due to J. Hutchinson. (The principle of this scheme first arose long ago, in the work of Poincaré and Klein, and corresponding illustrations using randomization are found in our *Fractal Geometry of Nature*.) Start with an "initiator" that is an arbitrary bounded set, for example is a point P_0. Denote the three similarities of the gasket by S_0, S_1 and S_2, and denote by $k(m)$ a random sequence of the digits 0, 1 and 2. Then define an "orbit," as made of the points $P_1 = S_{k(1)}(P_0)$, $P_2 = S_{k(2)}(P_1)$ and more generally $P_j = S_{k(j)}(P_{j-1})$. One finds that this orbit is "attracted" to the gasket, and that after a few stages it describes its shape very well.

Surprise: simple rules can generate rich structure

How did fractals come to play their role of "extracting order out of chaos?" To understand, one must go beyond simple shapes like the gasket or like the other fractals-to-be that mathematicians have first introduced as counter-examples. In these "old" shapes, indeed, what

Figure 1. The Sierpiński Gasket: Construction and Self-Similarity Properties, Static and Dynamic.

The four small diagrams show the point of departure of the construction, then its first three stages, while the large diagram shows an advanced stage. The basic step of the construction is to divide a given (black) triangle into four sub triangles, and then erase (whiten) the middle fourth. This step is first performed with a wholly black filled-in triangle of side 1, then with three remaining triangles of side 1/2, etc.

Perform a similarity (or more precisely a homothety), that is, an isotropic linear reduction whose ratio is 1/2, and whose fixed point is either of the three apexes of the triangle that circumscribes the gasket. It is obvious by examining the large advanced stage picture, that each of the three reduced gaskets is simply one third of the overall shape. For this reason, the fractal gasket is said to have three "self-similarity" properties.

As defined, these self-similarity properties seem "static" and "after-the-fact." However, it is also possible to reinterpret them as forming together a generalized dynamical system. In that case, the gasket becomes redefined as a dynamic "attractor" one gets out follows easily from what has been knowingly put in. The key to fractal geometry's effectiveness resides in a very surprising discovery the author has made, largely thanks to computer graphics.

The algorithms that generate the other fractals are typically so extraordinarily short, as to look positively dumb. This means they must be called "simple." Their fractal outputs, to the contrary, often appear to involve structures of great richness.

A priori, one would have expected that the construction of complex shapes would necessitate complex rules. Thus, fractal geometry is the study of geometric shapes that may seem chaotic, but are in fact perfectly orderly.

Let us, for the sake of contrast, comment on the examples of a related match between mathematics and the computer that arise in areas such as the study of water eddies and wakes. In these examples, the input in terms of reasoning or of number of lines of program is extremely complicated, perhaps more complicated even than the output. Therefore, one may argue that, overall, total complication does not increase in those examples, merely changes over from being purely conceptual to being partly visual. This change over is very important and very interesting, but fractal geometry gives us something very different.

What is the special feature that makes fractal geometry perform in such unusual manner? The answer is very simple: the algorithm involve "loops." That is, the basic instructions are simple, and their effects can be followed easily. But let these simple instructions be followed repeatedly. Unless one deals with the simplest old fractals (Cantor set or Sierpiński gasket), the process of iteration effectively builds up an increasingly complicated transform, whose effects the mind can follow less and less easily. Eventually, one reaches something that is "qualitatively" different from the original building block. One can say that the situation is a fulfillment of what is general is nothing but a dream: the hope of describing and explaining "chaotic" nature as the cumulation of many simple steps

Fractal geometry viewed as a language

"Mathematics *is* a language"" Josiah Willard Gibbs (speaking up at a Yale faculty meeting . . . on elective course requirements).

"Philosophy is written in this vast book — I mean the Universe — which stands forever open to our gaze, but cannot be read until we have learnt the language and become familiar with the characters in which it is written. It is written in the language of mathematics, and its characters are triangles, circles and other geometrical figures, without which it is humanly impossible to understand a single word of it; without which one wanders in vain through a dark labyrinth." (Galileo Galilei: *Il Saggiatore = The Assayer*, 1623)

"The language of mathematics reveals itself unreasonably effective in the natural sciences . . . , a wonderful gift which we neither understand nor deserve. We should be grateful for it and hope that it will remain valid in future research and that it will extend, for better or for worse, to our pleasure even though perhaps also to our bafflement, to wide branches of learning." (Eugene Wigner, 1960)

Inspired by the above quotes, the best is to call fractal geometry a new *geometric language*, which is geared towards the study of diverse aspects of diverse objects, either mathematical or natural, that are not smooth, but rough and fragmented to the same degree at all scales. Its history is interesting and curious. As is clearly indicated by such terms as "Cantor set," "Peano curve", "Sierpiński gasket", etc. . . . , several fractal "characters" date to 1875-1925. They count among the least complex (hence least beautiful), but they have seen previous use in other languages that have nothing to do with fractal geometry. As a language addressed to the goals discussed in this paper, fractal geometry was born with the first edition of our book *Les objets fractals*.

There is a profound historical irony in the fact that these old "characters" of the new geometry had been among the "monsters" to which we have referred earlier. The general monster is *not* scaling, and the fact that some monsters are scaling had not been singled out, because it was viewed as a "special" property, therefore one that is not very "interesting." In the mathematical culture of the century that ran from 1875 to 1975, special properties did not warrant investigation.

Clearly, the study of order in geometric chaos could not arise as a specialized object of study, and a term to denote it did not become indispensable, until we had performed two tasks: A) We saw that diverse rough patterns of nature — noise and turbulence and diverse geographical features — are geometrically scaling. Eventually we — and now many others — have identified geometric scaling in many other areas of nature, and then explored its consequences. B) We saw that the proper tool to tackle scaling in nature is suggested by some of the old "monsters." The old monsters themselves are not realistic models, but the construction of new fractals was immediately spurred by the new need, and more accurate models soon became available.

A language can be appreciated in diverse ways. For fractal geometry, the reasons can be sorted into five categories: artistic, mathematical, "historical," practical, and scientific. Let us discuss them in turn.

The language of fractals often creates plastic beauty

We use the term "geometry" in a very archaic sense, as involving concrete actual images. Lagrange and Laplace once boasted of the absence of any pictures in their works, and their lead was eventually followed almost universally. Fractal geometry is a reaction against the tide, and a first reason to appreciate fractal geometry, because the "characters" it adds to the "alphabet" Galileo had inherited from Euclid, often happen to be intrinsically attractive. Many have promptly been accepted as works of a new form of art. Some are "representational," in fact are surprisingly realistic "forgeries" of mountains, clouds or trees, while others are totally unreal and abstract. Yet all strike almost everyone in forceful, almost sensual, fashion. The artist, the child and the "man in the street" never seem to have seen enough, and they had never expected to receive anything of this sort from mathematics. Neither had the mathematician expected his field to interact with art in this way.

In any event, fractal geometry shows that there is an unexpected parallel to the above classic quote from Wigner: We have been fortunate to witness the revelation of the *"'unreasonable' and 'undeserved' effectiveness of mathematics as a source of enjoyable form."*

Fractal geometry and mathematics

One can also appreciate the language of fractals because it happens to have led to new mathematics that several groups of specialists find attractive. To the layman, fractal art tends to seem simply magical, but no mathematician can fail to try and understand its structure and its meaning. The remarkable aspects of recent events is that much of this mathematics, had its origin been hidden, could have passed as "pure," in other words, as absolutely self-referential.

Many of the early pictures of fractals have mostly served to "visualize" facts that had been obtained previously by abstract "pure mathematical" thought. For example, one who examines the computer-generated pictures of Julia sets and Fatou domains, which are mathematical objects that concern the dynamics of "rational of iteration," cannot help being filled with deep humility and boundless admiration for the creative powers Pierre Fatou and Gaston Julia had exhibited in 1917-1919. He may also be tempted to echo Georg Cantor's assertion "that the essence of mathematics resides in its freedom." Such arrogance has indeed characterized much of mathematics since the period 1925-1975, but an examination of further examples of fractal art necessarily brings it to an end. To many mathematicians, the newly opened possibility of playing with pictures interactively, has turned out to reveal a new mine of purely mathematical questions and of conjectures, of isolated problems and of whole theories. To take an example, examination of the Mandelbrot set leads to many conjectures which were simple to state, but then proved very hard to crack. To mathematicians, their being difficult and slow to develop does not make them any less fascinating, because a host of intrinsically interesting "side-results" have been obtained in their study.

Herein hangs a tale. Pure mathematics certainly does exist as one of the remarkable activities of Man, it certainly is different in spirit from the art of creating pictures by numerical manipulation, and it has indeed proven that it can thrive in splendid isolation — at least over some brief periods. Nevertheless, the interaction between art, mathematics, and fractals confirms what is suggested by almost all earlier experiences. Over the long haul mathematics gains by not attempting to destroy the "organic" unity that appears to exist between seemingly disparate but equally worthy activities of Man, the abstract and the intuitive.

Nevertheless, fractal geometry is not (at least as of today) a branch of mathematics like, for example, the theory of measure and integration. It fails to have a clean definition and unified tools, and it fails to be more or less self contained. Calling it "fractal geometry of nature and of chaos," immediately explains why it is more like the theory of probability. Both thrive only when they are not over-defined, when they do not mind using mismatched tools, and when they overlap heavily with many neighboring endeavors. Moreover, the comparison with probability is not meant to imply a comparable level of development. While the fractalists have some reason for modest pride, their mathematical language keeps continuing to evolve and to expand with each new use. To find that a new use require the improvement of some basic point is not an exception, rather an everyday experience.

Some great old "problems that pose themselves"

There is a third reason to appreciate fractal geometry, whose validity is totally independent of the earlier two. This new reason involves the antiquity and the prevalence of the natural phenomena on which it has made a dent, either via descriptions that suffice to the engineer, or via theories that satisfy the sophisticate.

The great Henri Poincaré drew, long ago, a distinction between those problems that a scientist chooses to pose, and those problems that pose themselves. (He then went on to praise Paul Painlevé for having identified a problem that nature has been attempting to pose, and for having tackled it.) To sort out all problems of Nature in this fashion is not an assignment that a prudent scientist will want to undertake. Besides, a scientist gains no merit by contemplating problems which his tools are too feeble to solve. Nevertheless, Poincaré's distinction is intuitively an essential one. Does a new scientific enterprise contrive new problems for the pleasure of solving them easily? Or does it, however short it may remain from achieving perfection, contribute to sharpening and understanding some problems that are very old.

First example. To try and answer these questions in the case of fractal geometry, the best is, of course, to first allude to the mountains, the clouds, the water eddies and the trees. To the highest degree, the problems raised by their geometry deserve to be described as being "problems that pose themselves." Man-the-artist must have been pondering them forever, but Man-the-scientist did not know how to start tackling them. (At this point of the argument, an etymological digression comes to mind.

The root of *geometry* is the Greek $\gamma\epsilon\omega\mu\epsilon\tau\rho\iota\alpha$. The customary translation, "measure of the Earth," used to suggest to us that "geometry" had once denoted the problems we now tackle by fractals. But we are no longer sure: did it denote a far-reaching "measure of the Earth," or a practical "measure of the land" (as claimed in Eric Partridge's *Origins*). The *Oxford English Dictionary* points out that "in early quotations, geometry is chiefly regarded as a practical art of measuring and planning, and is mainly associated with architecture.")

Second example. We continue with quotes from the King James Version of the Bible.

. . . were all the fountains of the great deep broken up, and the windows of heaven were opened. And the rain was upon the earth forty days and forty nights. *Genesis, 6*, 11-12.

. . . there came seven years of great plenty throughout the land of Egypt. And there shall arise after them seven years of famine . . . *Genesis, 41*, 29-30.

Awe of the actual unpredictability of the weather obviously permeates the stories of Noah's Flood, and of Joseph son of Jacob, the interpreter of Pharaoh's dream of the Seven Fat and Seven Lean Cows. More down-to-earth records confirm that such occurrences are in fact common in meteorology. On the other hand, none of the standard tools of modeling can account for these symptoms. Arguing that this failure could not be "fixed" by small changes led us, many years ago, to draw a certain sharp distinction, to coin for it a flippant but immediately useful terminology, and to introduce very different tools that eventually become integrated in fractal geometry.

When a natural phenomenon is such that its action is felt much of the time, yet most of its effects concentrate "oligopolistically" in one or a few largest events, the phenomenon is now said to obey the *Noah Effect*. A phenomenon is said to obey the *Joseph Effect* when it involves "trends" of arbitrary slowness, whose cumulative effect exceeds the effect of even the largest addend.

We cannot dwell upon our studies of these two Effects, beyond saying that the tools we used in these studies were eventually redesigned to yield our fractal models of mountains and of clouds.

Third example. It is "turbulence." To announce that at long last "a cure" for it has been found seems a way of securing publication and an audience, but there is no cure as yet, only small advances raising big new questions. Turbulence can be taken as a prototype of the fractal phenomena that are old and interesting, and at the same time resistant to analysis. Several of the decisive early steps towards fractal geometry, such as the introduction of multifractal measures, were taken in 1968-1976 when we were working on turbulence. Multifractal and fractal tools have contributed their share to descriptions and understanding, but turbulence has not been explained, yet.

Fourth example. It is far less widely known than the preceding three, but far more relevant to the topic of the present meeting. It concerns *fractal growth* models and *fractal aggregates*. The discovery of diffusion limited aggregation (DLA) by Witten and Sander has initiated one of the most surprising and challenging quests of the last ten years in condensed matter physics. A complete understanding of the fractal and multifractal properties of such aggregates remains an open and intriguing problem, to which we shall return momentarily. Yet, one who sees a fractal aggregate for the first time is likely to experience a feeling of "déjà-vu".

Fractals and the hard-pressed engineer

Yet another reason to appreciate the language of fractal geometry is in some way a restatement of the third. The new language promises to be effective in engineering. Self-styled sophisticates tend to either forget or spurn the needs of the practical man. One reason is that he does not have the luxury of waiting until the phenomena he chooses, or is asked to try and control have been explained to the satisfaction of the sophisticates. Instead, he finds himself lost in Galileo's "dark labyrinth," where he does not even know what signs to look for or what to measure.

The list of investigations where fractals matter already or promise to matter very soon from the viewpoint of phenomenology and of engineering is already long, but to comment in turn on each "case" would be tedious and would generate little light. Let us, instead, give a few typical examples.

Rock-bottom, a prerequisite of engineering, and also a continuing goal of science, is to *describe* nature quantitatively. But everyone who has tried knows that *to see* is a skill one must learn, and that one must learn what to measure. All too many disciplines harbor the strong wish of becoming quantitative, but do not know even how to begin. One standard way is to ask new questions for known answers, that is to borrow procedures from disciplines that have already reached a quantitative stage, and to hold on to these procedures if they appear to be effective. One finds that procedures one could borrow are not particularly numerous. While the diversity of nature appears to be without bound, the number of techniques one can use to grasp nature is extremely small and increases very rarely. Therefore, the enthusiasm usually generated by the birth of a new technique and the desire to test it more widely is healthy, and must not be disparaged. In a number of notable cases, the new fractal additions to Galileo's geometric alphabet prove to be of great help in efforts to see and to measure. Let us examine a few cases.

Viscous flow through porous media, for example the flow of water pushing oil, has proven recently to admit to several regimes, one of which is a "front," which is an effective and desirable configuration, and another one is "fractal fingering," which is undesirable. After this range of possibilities had become known, a colleague of ours saw a very old article on viscous flow. Next to a photograph that could have been taken yesterday, a diagram meant to summarize what the original photographer had seen in his work. Unfortunately, what he had seen turned out to lead nowhere, but what he has smoothed away turned out to prove important, and fractal.

Our second example is from hydrology. There is no question that the design of aqueducts and of dams involves many aspects of the science of materials. But what about the variability of the water discharge in rivers? Even if a full climatological explanation were available for the long-run component of this very erratic process, tasks have been assigned to different professions in such a way that the availability, or lack of availability, of a climatological explanation cannot possibly matter to the water resources community. Yet, when, starting in 1963, we advanced a model of the long run persistence in river discharges, we found that explanation was perceived as mattering very much. Our model having an infinite memory span. The first and most frequent questions were the following: "Why choose so peculiar a model?" "Has this model already been seasoned in the usual way, by being used in physics?" and "What is the climatological explanation of this model?" In the water resources community, the most quantitatively inclined practitioners seemed intimidated by assertions of the primacy of explanation over everything else, afraid perhaps of hearing someone thunder "But where is the science behind what you do?" We argued that it is best to perform each task in its time. The irony is, of course, that — within a few years — our model ceased altogether to be "peculiar," because close counterparts were discovered, and soon adopted, in many chapters of "mainstream" physics. These counterparts , as well as our model, have become building blocks of fractal geometry.

Our third example is from economics. To elaborate further upon the difference we see between the roles of fractals in engineering and in science, let us mention that security and commodity prices had been in the early sixties the topic of the first descriptive account that was to be later counted as fractal. Starting with the assumption that price is a continuous and differentiable function of time. We claimed not only that it is not obvious, not only that it is contrary to the evidence, but that it is in fact contrary to what should be. The reason is that a competitive price should respond, in part to changes in anticipation, which can be subject to arbitrarily large discontinuities. (The event of October 19, 1987 comes first to mind in 1988, but in the early 1960's such examples were viewed as things of the past.) On the basis of discontinuity combined with suitable self-similarity, we proposed a model of price variation, which had to incorporate the property of infinite variance of price increments. Our study of prices kept eliciting the comments already mentioned in the context of hydrology, the third one being phrased as follows: "Your models look fine, but how do you relate them to economic theory?" In moments of irritation, we are quoted as responding "There is, as yet, no explanation for these findings; in fact *no* explanation could reasonably be expected to come from existing economic theory. After all, this theory has been growing for well over a century, and has yet to predict anything."

The preceding section shows how difficult and unpopular is the task of defending the worth of scientific investigations that have not "risen above" phenomenology. This difficulty is not new, in fact is very familiar in everyday life, where one can and must deal honestly with imperfection, while both preserving and keeping in check the dream of a More Heavenly City. In the context of fractal geometry, the Heavenly City of Explanation has already been reached in several cases, of which notable examples will be given. Everyone should rejoice, but it must immediately be acknowledged that other cases remain more like everyday life. The scientific aspects of fractals are best discussed in two groups: the most highly developed, and the most challenging ones.

The role of fractals in the theoretical sciences

The most highly developed examples of this role divide naturally into those which involve chance, and those which do not. The former are best exemplified by a chapter of statistical physics, called percolation theory, in which fractals have risen highest above rock-bottom phenomenology. In this chapter, the fractal description is admirably complete, and the physics have been shown to be ruled by geometry, via a small number of quantities, each of them the fractal dimension of some specified portion of a geometric shape called critical percolation cluster. Furthermore, the basic dimensions have been deduced from basic physics, and some even turn out to be rational numbers! The reduction of physics to geometry being one of the basic goals of physics, the role of fractals in percolation theory is close to perfection — even though the mathematician will interject that many basic facts that the physicist holds true have not been proven in full rigor, at this point. Most unfortunately, however, percolation theory is well outside the main highways of mathematical physics, so that it is best to move on.

A second source of completely understood random fractals is found in probability theory, namely in the sets or measures that illustrate limit theorems concerning sums or products of random effects. Brownian motion is the prime example, one that has preceded fractal geometry by decades but has been incorporated in its fold. The Brown surfaces we have used as models of relief are another example. Self-avoiding random walk also shows every sign of being a fractal, though (again) the fact has not been proven in full rigor. The basic multifractal measures are obtained as limits of products of random factors.

Among non random fractals of immediate relevance to physics, the best understood ones result from the fact that "basins of attraction" are usually bounded by fractal sets. Let us give an intuitive example, and then amplify this statement. Since a drop of water that falls on the United States and does not evaporate will eventually end in either the Atlantic or the Pacific, one can say that each of the two Oceans has a basin of attraction. The two basins are bounded by the Continental Divide, which our fractal model of relief happens to represent usefully by a fractal curve. More formally, consider a dynamical system that eventually converges to either of several limit states. Each limit state has a basin of attraction, and the typical situation is when the boundaries between the basins are fractal sets. The first example was advanced by Pierre Fatou, in the work that was already mentioned when we discussed the impact of fractals upon mathematics. In this case, which is why all the basins of attraction taken together are now called Fatou set, the boundaries of these basins being called Julia sets. The broad impact of the old Fatou and Julia work was felt only after it had been made part of fractal geometry. From the viewpoint of physics, however, this work lack direct relevance because it concerns rational maps in discrete time. The most important recent development may well be the proof that basin boundaries are also fractal in more general maps in continuous time.

Fractals and the great equations of mathematical physics

One can, last but not least, appreciate the language of fractals because it has already been used by master physicists and other scientists to produce beautiful works that add to our understanding of how the world is put together, and in particular of our understanding of the extent to which there is truth in the widely held notion that geometry rules physics.

What is generally perceived as the highest level of natural science is the study of the great old equations of mathematical physics: those of Euler, Laplace, Fourier and Navier-Stokes. The fact that these equations are *differential* implies a degree of smoothness that may seem a priori to forbid any connection with the rough and fragmented world of fractals. But it turns out that connections do exist, and that they are very important, even though some of them may have raised new questions in numbers exceeding the old questions they have already solved.

Our own contribution, which was the first interplay of fractals and differential equations, concerns the equations of fluid motion. We started with the old idea of Oseen, developed by J. Leray, that "turbulence" is the name one gives to the effects of the singularities of Navier-Stokes equations. Granting this notion has allowed us to transform our fractal or multifractal models of turbulence into conjectures concerning the singularities of the equations. In specific mathematical terms, we introduced Hausdorff dimension as a new concern and a new tool in the study of the Navier-Stokes equations. The tool has proven attractive and the study to be fruitful, but by no means easy.

In the study of the Fourier equation, fractals have found a very attractive role in the notion of fractal fronts and of fractal foams of diffusion. The description we like best concerns a finite and discrete triangular lattice with a (hexagonal) ball placed on each lattice site. Let the abscissas of the balls run from $-x_{max}$ to $+x_{max}$, and start with the initial conditions where a ball is the color of sand if its abscissa is negative, and is blue if the abscissa is positive. At each discrete instant of time, let "couples" of neighboring balls be chosen at random, independently of each other, and let their colors be interchanged. (The probability of being chosen is taken to be small, and if chance wants to choose a given ball as part of more than one "couple," this ball can be left alone.) After a while, the relative proportion of blue balls will become a function of x, and the expected relative proportion is well-known from the theory of diffusion of heat. Expectations, however, give an incomplete view of reality. What about the precise "front" between the two colors? To define a front, reverse the color in the sand islands entirely surrounded by blue sea, and in the blue lakes entirely surrounded by sand-colored beach. This leaves us with a blue sea and a sand beach, separated by a shore line. Well: this shoreline happens to be a fractal curve, and its dimensions happens to be $D = 7/8$. To be more precise, it does not become a fractal curve until the sites are downscaled to infinitesimal sizes. As it wanders back and forth, this shoreline defines pen-insulas that are attached to the beach by a single linking site. Imagine that these linking sites' colors are reversed. This will create a messy "foam" separated from both beach and sea by wiggly curves: both wiggly boundaries happen to be fractal curves, and their dimension happens to be $D = 4/3$. Higher-dimensional space diffusion creates an even more significant "diffusion" foam. The proofs of the above assertions belong to the theory of percolation; hence, they are rigorous by the standards of physics, but continue to open purely mathematical problems.

The last but not least of the links between fractals and mainstream mathematical physics is raised by DLA (diffusion limited aggregation) and its many variants. It concerns the Laplace equations with interplay between, on the one hand, the solution of the equation at time t, and, on the other hand the displacement of the boundaries between times t and $t+1$. The body of this book, however, contains so much about DLA, that it is hardly worth dwelling on it. This new problem promises to be difficult, and to stay with us for a long time. A conservative may even claim that it should have been mentioned in an earlier section concerned with phenomenology, but this is a minor issue. It is, we think, a feather in a cap of fractal geometry that it has created in advance the environment and the tools that had allowed the study of DLA to move forth as quickly as it had done.

Reprinted with permission from *Scientific American* **256** (1), L.M. Sander, "Fractal Growth," pp. 94–100, ©1987.

Fractal Growth

Growth in nature can produce the sprawling, tenuous patterns called fractals. One kind of fractal growth explains such diverse phenomena as how some solids crystallize and how air bubbles move in fluids

by Leonard M. Sander

Investigators who study bulk matter face a problem of bewildering complexity. Each macroscopic bit of the world contains gigantic numbers of atoms and molecules that are often arranged in complicated, disorderly patterns. In the case of a perfect crystal or a smoothly flowing liquid, the pattern is uniform on large scales. Even though a great deal can be understood about such systems, the vast majority of complicated natural phenomena—such as turbulent flows of fluid or air, the accretion of metal particles in an electrolytic bath and the formation of mountain ranges—have virtually defied understanding.

In the past 10 years both scientists and mathematicians have made much progress toward gaining such comprehension. Central to many of the new insights is the revolutionary concept of a fractal, a term coined by Benoit B. Mandelbrot of IBM's Thomas J. Watson Research Center in Yorktown Heights, N.Y. A fractal is an object with a sprawling, tenuous pattern. As the pattern is magnified it reveals repetitive levels of detail, so that similar structure exists on all scales. A fractal might, for example, look the same whether viewed on the scale of a meter, a millimeter or a micrometer (a millionth of a meter). Mandelbrot pointed out that many disorderly objects in nature have this property.

An increasing amount of evidence suggests that nature's love of fractal shapes is a deep one. Fractals known as percolation clusters have been identified with the pattern of a fluid flowing through a solid matrix, such as water seeping through the soil or coffee through ground coffee beans. Soot, colloids and some polymers seem to be fractals. Fractals also occur in the motion of air bubbles in oil, the growth of some crystals and the behavior of electrical discharges resembling lightning bolts. The random patterns of clouds and coastlines are almost certainly fractals as well.

As empirical evidence for the existence of fractals in nature mounts, investigators have begun to explore the question of how fractals form. In 1981 Thomas A. Witten III of the Exxon Research and Engineering Company and I proposed a mechanism for fractal growth that we call diffusion-limited aggregation. According to our model, a particular kind of fractal can result from a disorderly and irreversible growth process. The theory is attractive for two reasons. First, it is conceptually simple and easily modeled on a computer. More important, it seems to explain how a variety of real fractals form.

What are the properties of a fractal? Abstract discussions of objects now called fractals were given long before Mandelbrot by other mathematicians, who viewed them as "monsters" of only academic interest. Yet a typical fractal reminds people more of a snowflake than a monster. The reason is found in the repetition of a pattern. Each unit of the fractal shown at the bottom of the illustration on page 96, for instance, consists of five identical subunits. Five of the large units can then be put together to make a still larger unit, and so on. Every generation contains holes scaled to the size of that particular generation. The pattern is also scale-invariant: at each stage any part of the pattern that has one-third the diameter of the whole looks exactly like the whole. Scale-invariance is a "symmetry" of fractals. Just as round objects are symmetric under rotations, fractals are symmetric under dilations, or changes of scale.

It is useful to have a quantitative measure of how a fractal scales. This is given by a number called the fractal dimension. A fractal dimension, as opposed to an ordinary dimension, is expressed not as a whole number but as a fraction. The fractal under discussion, for example, has a dimension of 1.46: it is intermediate between a one-dimensional straight line and a two-dimensional plane. The more nearly a flat fractal object fills a plane, the closer its dimension approaches 2. The pattern shown in the computer image on the opposite page, which was formed by the diffusion-limited-aggregation process, has a measured fractal dimension of 1.71. The pattern is scale-invariant in a statistical sense. One property of that fractal—and in fact of all fractals—is that its density decreases as its size increases.

The fractal dimension of a physical object is a "universal" property: it is independent of many details of how the object is formed. The fractal dimension and other universal properties are related to behavior at large scales, where the particular details average out. As a consequence, a simple model that neglects most of the complexity of a real system should nonetheless describe the scaling properties of the system correctly.

The importance of the diffusion-limited-aggregation model is that it

FRACTAL pattern was produced by computer simulation of a process called diffusion-limited aggregation. Some 50,000 "particles" were released one at a time from an area outside the illustration and allowed to wander toward the origin. As they stuck to one another a cluster was formed, and then it grew. The image was made by Paul Meakin of E. I. du Pont de Nemours & Company, Inc.

1250 LATTICE UNITS

shows a relation between fractals and growth. There are many ways that objects in nature can grow. A perfect crystal, for instance, grows near equilibrium: it "tries" many configurations until it finds the state with the most stable structure. When a molecule is added to the growing crystal, it must in general search over many possible sites before it finally sticks in a favorable place. An equilibrium crystal forms slowly and is subject to constant rearrangement. Most real processes of growth, however, do not have the luxury of time. All biological life, for example, is out of equilibrium. The fractals I want to discuss grow far from equilibrium. (Some fractals grow near equilibrium, but that is beyond the scope of this article.)

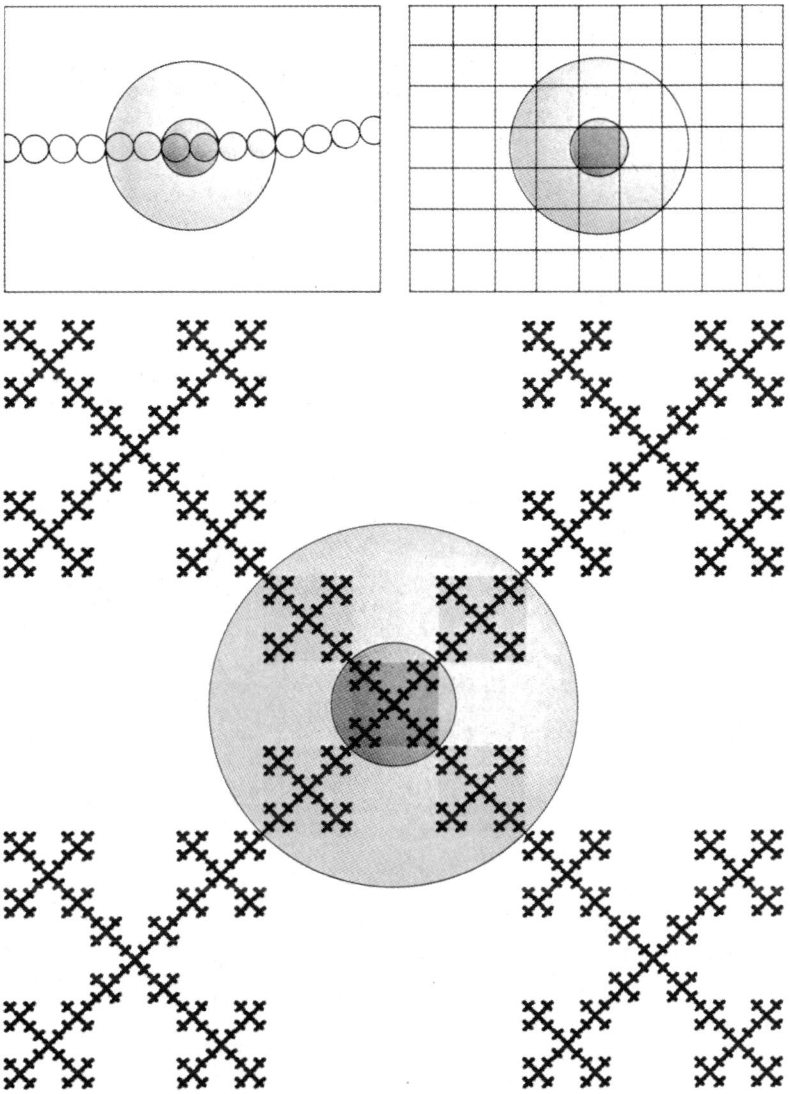

FRACTAL DIMENSION differs from an ordinary dimension in that it is not expressed as a whole number but as a fraction. To determine the precise fractal dimension of an object one counts the average number N of fundamental units of repetition found within a sphere of a certain radius r centered somewhere on the object. According to Euclidean geometry, the number of fundamental units then equals a constant C multiplied by the radius raised to the value of the dimension D ($N = C \cdot r^D$). In the case of a line the dimension is, of course, 1: tripling the radius of the sphere triples the number of units inside (*top left*). For ordinary (nonfractal) bulk matter in two dimensions, tripling the radius of the sphere multiplies the number of units by 9 (*top right*). For a fractal of dimension 1.46 (*bottom*), in contrast, tripling the radius multiplies the number of units by 5. In other words, the number of units grows faster than it does in the case of a line but not as fast as in the case of ordinary bulk matter. In this sense the object is intermediate between a line and a plane. The fractal shown here was invented by Tamás Vicsek of the Research Institute for Technical Physics of the Hungarian Academy of Sciences in Budapest.

Imagine growing a cluster by adding one particle at a time so that as each particle comes in contact with the growing object, it sticks and never tries another site, that is, it simply stays put. Such a process is called aggregation. It is an extreme example of a nonequilibrium process, because no rearrangement takes place at all. Now suppose the particles diffuse to the cluster by means of a random walk: a sequence of steps whose magnitude and direction are determined by chance. (In a one-dimensional version of a random walk, for instance, a person flips a coin and takes one step forward if the result is heads and one step backward if the result is tails.) The aggregation of particles by means of random walks is what Witten and I called diffusion-limited aggregation.

Small clusters are quite readily "grown" on a personal computer. A cluster is started by placing a particle at an origin. Then another particle is released some distance away and allowed to take one random step after another until it approaches within a particle diameter of the first. When the second particle sticks, another is released at a random location far from the aggregate, and so on. The simulations suggest that diffusion-limited-aggregation clusters are fractals.

Although diffusion-limited aggregation is simple to describe and to simulate, the process is still not well understood at a deeper level. Why, for example, does diffusion-limited aggregation give rise to fractals rather than to, say, amorphous blobs with no symmetry at all? Why are loops rarely formed? How does the fractal dimension depend on the dimension of space? Answers to these questions remain at large, and they pose a remarkable problem to the theoretical physicist because none of the ordinary mathematical tools seems to work when applied to them.

It is possible, however, to understand in a qualitative way some important features of the process. Imagine beginning with a smooth cluster onto which diffusing particles aggregate. When the cluster is small, several incoming particles may stick—purely by chance—on one part of the surface. In other words, tiny bumps and holes will form on the surface owing to "noise," or random statistics, in the behavior of the particles.

Once the surface has bumps and holes, the bumps will grow much faster than the holes. The reason is that a random-walking particle, moving tortuously from the exterior, will probably stick at or near the peak of a bump; it will almost certainly stick on the side before it could fall into a hole. Since the particle sticks near the peak, the bump grows even steeper; the filling of a hole becomes even more unlikely. As a result a cluster that is initially slightly distorted will become even more distorted, an effect called a growth instability. Eventually the growth and splitting of the protruding ends presumably give rise to a fractal. Even though the details of the process remain unknown, it is the interplay of noise and growth that is the source of the complexity and richness of diffusion-limited-aggregation clusters.

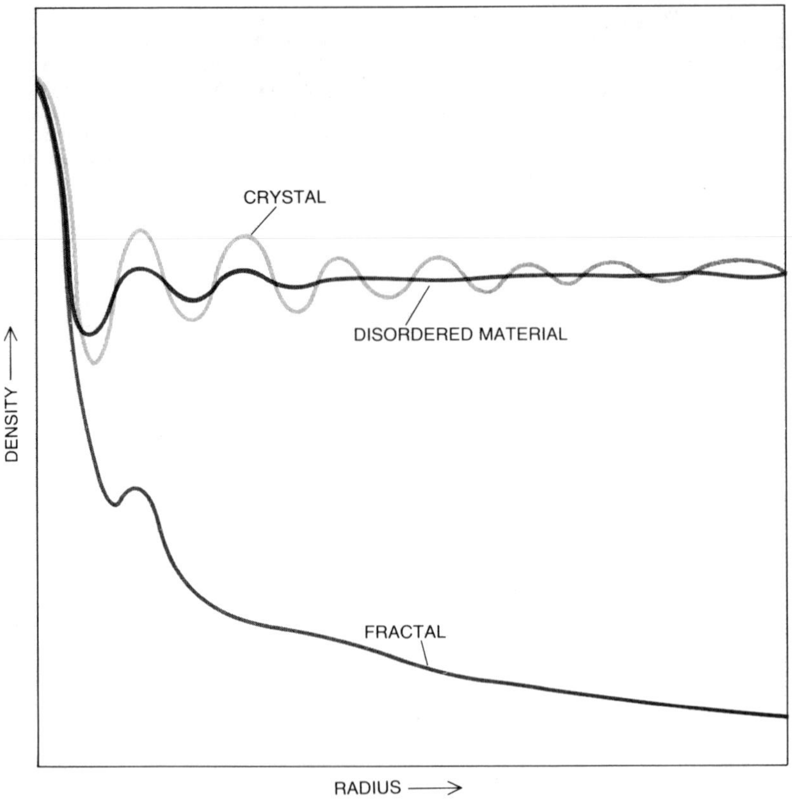

DENSITY OF A FRACTAL decreases with increasing size, as shown here. The densities of both ordered crystals and amorphous blobs, in contrast, approach fixed constants.

In the past five years diffusion-limited aggregation has been the subject of extensive inquiry. A great deal of excitement springs from the fact that the model seems to describe the real world: actual particles often do wander to sites, where they stick. Robert M. Brady and Robin C. Ball of the University of Cambridge pointed out in 1984 that diffusion-limited aggregation is, for instance, a reasonable idealization of what happens when a metal is deposited from an electrolytic solution of diffusing ions. Although the details of what happens when an ion sticks to the deposit are surely quite different from what happens in a computer simulation, they do not seem to affect the overall resulting shape. Nor do the details affect the resulting fractal dimension.

A deposit of zinc metal produced in an electrolytic cell [*top left in illustration on page 100*], for example, bears a striking resemblance to the computer-generated fractal pattern on page 95. The zinc deposit has a measured dimension of 1.7, which, taking experimental error into account, agrees with the measured dimension of the computer-generated fractal (1.71). The agreement is a remarkable instance of universality and scale-invariance: the computer simulation used about 50,000 points, whereas the number of zinc atoms in the deposit is enormously large, almost a billion billion.

In fact, varying the rules for the computer simulation reveals several types of universality. Suppose, for instance, a particle sometimes bounces off the aggregate instead of sticking. That rule, which is a simple representation of one of the many complications that might develop in a real physical situation, turns out to lead to a thickening of the branches of the aggregate, but it does not change the fractal dimension.

It is certainly easy to believe the deposition of a metal onto an electrode is described by diffusion-limited aggregation. Remarkably, however, the model also appears to account for a wide range of phenomena. One phenomenon involves an experimental apparatus called a Hele-Shaw cell, adapted from the work of a 19th-century British naval engineer, Henry S. Hele-Shaw. The cell consists of a viscous fluid, such as glycerine, confined between two parallel plates. When a less viscous fluid, such as air, is injected into the middle, the glycerine is displaced. An air bubble forms, sprouting a number of "fingers," or protruding ends [*top right in illustration on page 100*]. The phenomenon is called, appropriately enough, viscous fingering.

Viscous fingering is of practical interest because it also occurs when water is injected into the middle of an oil field during enhanced recovery operations. The efficiency of the recovery is greatly decreased by the fingering pattern. Unless special techniques are employed, one can displace only a small amount of the viscous oil toward the wells at the edge of the field.

The patterns that result from viscous fingering show a strong similarity to computer-generated images of diffusion-limited-aggregation clusters. Why is this so? The answer was provided recently by Lincoln Paterson of the Commonwealth Scientific and Industrial Research Organization in Australia. Paterson points out that the processes in both diffusion-limited aggregation and viscous fingering are the same in principle. In the former, growth is due to the net inward flow of random-walking particles. The flow arises from the fact that particles are more likely to wander in from densely populated regions outside the cluster than from sparsely populated regions. Specifically, the flow is proportional to the rate of change in the populations outside the aggregate.

In the case of viscous fingering, the fluid pressure in the glycerine is analogous to the populations of particles. The pressure is highest at the boundary between the air bubble and the glycerine. The pressure is relieved by the flow of glycerine away from the bubble. The rate of flow is proportional to the rate of change in the pressure at points outside the bubble. The fingers grow because fluid flows away from them most easily. Since the boundary moves when the fluid flows away, the tips grow longer. As a result there is a growth instability similar to that of diffusion-limited aggregation.

Yet another system has been analyzed in much the same way. When a voltage is applied to an electrode that touches a piece of photographic emulsion or a fine powder spread on the surface of an insulator, a jagged, stringy electrical-discharge pattern resembling lightning results [*bottom left in illustration on page 100*]. The pattern is called a Lichtenberg figure, after the 18th-century German physicist Georg Christoph Lichtenberg. In 1984 a

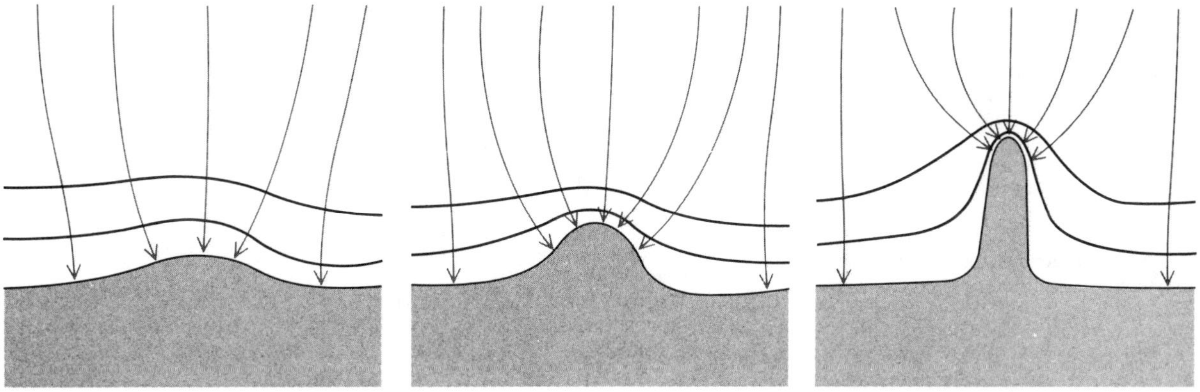

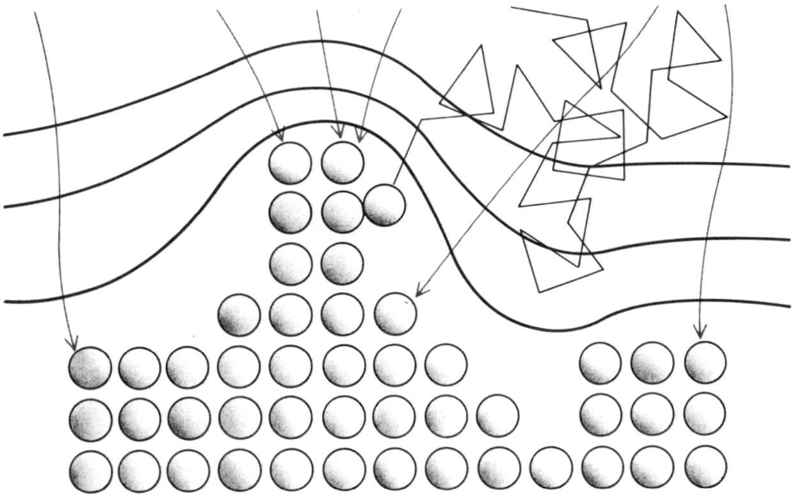

GROWTH OF FRACTALS by diffusion-limited aggregation is illustrated schematically. Typically one begins with a smooth cluster onto which diffusing particles aggregate. Tiny bumps and holes form on the surface owing to "noise," or random statistics, of the incoming particles (*left*). The black line shows the random path of an incoming particle, the colored lines are contours of constant average density of particles and the gray lines are streamlines of average flow. Once the surface develops bumps and holes, the bumps will grow faster than the holes (*top*). The reason is that an incoming particle, in its tortuous path from the exterior, is quite likely to stick at or near the peak of a bump; before it can wander deep into a hole it will almost certainly stick on the side. The sticking of the particle near the peak of the bump makes the bump grow even steeper, and so the filling of a hole becomes even more unlikely: a smooth cluster is quickly distorted.

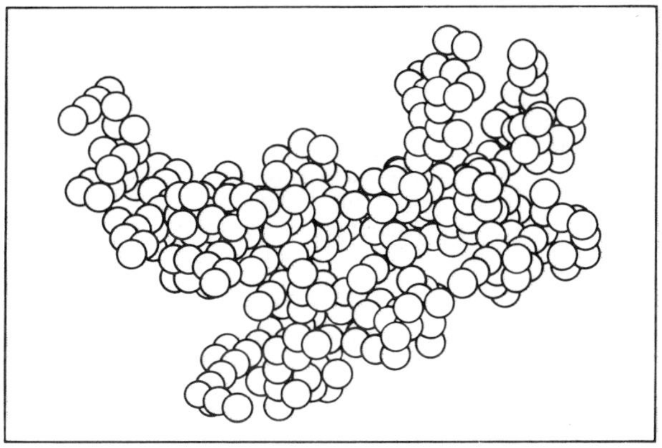

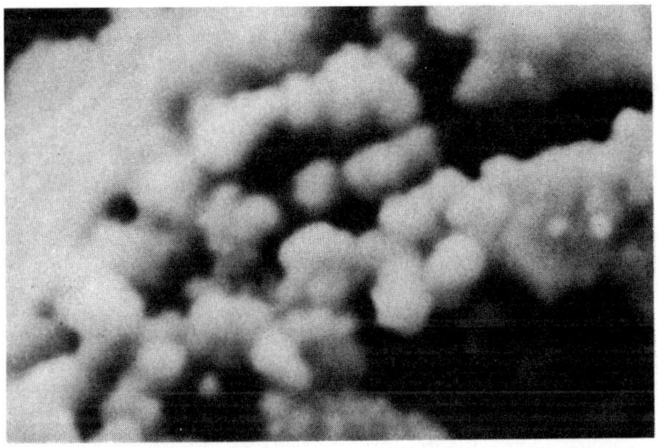

DIFFUSION-LIMITED AGGREGATION, simulated in a three-dimensional space by computer, gives rise to a fractal with a dimension of 2.4 (*left*). The pattern shows a close similarity to that of a copper cluster (*right*), which has the same dimension. The computer image was generated by Roy Richter of the General Motors Research Laboratories, and the copper cluster, which was deposited from a solution of copper sulfate, was prepared by Nancy Hecker and David G. Grier of the University of Michigan.

group of investigators at Brown, Boveri & Company, Limited, in Switzerland reported that diffusion-limited aggregation seems to underlie the growth of Lichtenberg figures.

Suppose the applied voltage is initially high enough near the electrode to damage a small region of the emulsion and create a channel of electrical conduction. Outside the channel there is an electric field; the strength of the field is given by the rate of change in the voltage at various points in the material. The Brown, Boveri group made the reasonable assumption that the growth of the channel will most likely occur where the electric field is strongest, which turns out to be at the sharp tips of the discharge. As a result the tips grow longer and proliferate: fractal growth takes place.

The common framework that unites the deposition of a metal on an electrode, the fingering of a viscous fluid and the formation of a Lichtenberg figure is most easily expressed in the abstract language of partial differential equations. It is not difficult, however, to gain some appreciation of the underlying effect by making a simple analogy to a rubber sheet, taut along all four edges and pressed down in the middle by a growing fractal.

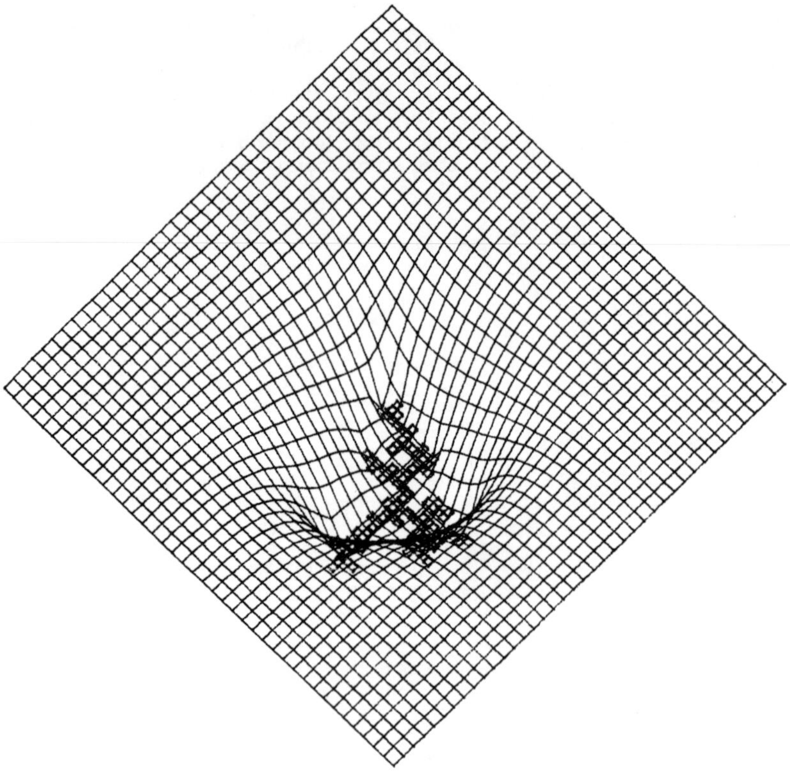

STRETCHED RUBBER SHEET, taut along all four edges and pressed down in the middle by a growing fractal, serves as a simple model of diffusion-limited aggregation. The fractal grows fastest at the places where the rubber sheet descends most steeply: the tips of the fractal. At the next stage still sharper tips will press on the sheet, and so on.

Functions such as the probability of where a particle will randomly walk, the pressure in a Hele-Shaw cell and the voltage near a breakdown channel are all examples of harmonic functions, which are solutions to a set of partial differential equations. A harmonic function has a net zero curvature: if it curves up in one direction, it curves down in the perpendicular direction (somewhat like a horse saddle). The rubber sheet also has zero curvature. Thus the height of the sheet can be thought of as a mapping of probability, pressure or voltage, and its slope at the edge of the fractal is the growth rate. The steepest slopes occur near the sharp tips, which grow preferentially. At the next stage still sharper tips press on the sheet, and so on.

It is tempting to speculate how far such an analysis can be pushed. For example, the random branching patterns of blood vessels, air passages to the lungs and coral reefs are certainly reminiscent of fractal patterns formed during diffusion-limited aggregation. Although some investigators have attempted to model the growth of these processes, to the best of my knowledge none of the approaches has explicitly invoked fractal geometry. Whether it will prove fruitful to analyze some kinds of biological growth in similar terms remains to be seen.

The model of diffusion-limited aggregation has also been used to describe other physical systems, such as surface crystallization of amorphous films. In addition a generalization of the model, called cluster-cluster aggregation, has been shown to describe colloids and aerosols such as soot. In this process, which was proposed by Paul Meakin of E. I. du Pont de Nemours & Company, Inc., and Max Kolb, Rémi Jullien and Robert Botet of the University of Paris at Orsay, many clusters can form and can themselves move and combine. In short, aggregation models have been extremely useful in describing physical systems.

At the same time I must also point out that fractals certainly do not account for all sprawling patterns found in nature. Snowflakes, for instance, are probably not fractals. Snowflakes are complex, to be sure, but they possess much more evident symmetry than diffusion-limited-aggregation clusters, and as such they are part of a family of crystals called dendrites. The beautiful macroscopic structure of a snowflake reflects the underlying microscopic anisotropy of the hexagonal lattices in which its atoms are arranged. One might then wonder why zinc, which also has a hexagonal form, deposits itself as a fractal in an electrolytic cell [*top left in illustration on next page*]. The answer is that the growth, albeit out of equilibrium, is so slow that the splitting of the tips wipes out the lattice anisotropy. Interestingly enough, when the growth rate is increased by raising the voltage of the cell, the anisotropy makes itself felt and a dendritic, or snowflakelike, pattern results [*bottom right in illustration on next page*]. The transition from fractal patterns to dendritic patterns is being explored by several groups of investigators.

So far I have concentrated on a particular growth process, diffusion-limited aggregation, that produces fractals. Can the knowledge be put to good use? In particular, do the scaling properties I have discussed lead to a useful understanding of physical characteristics of clusters other than their geometry?

Early signs indicate that they do. Over the past several years Raoul Kopelman and his colleagues at the University of Michigan have, for example, explored various chemical reactions that take place on percolation clusters (which are fractals formed in equilibrium). The investigators have shown that if the reaction is confined to the cluster, it acts oddly. Unlike a typical reaction, which occurs at a constant rate, the reaction rate on the percolation cluster appears to depend on time. The basic reason is that chemical species wandering on a fractal do not diffuse nearly as efficiently as they do in free space. They are hampered in finding one another because they are trapped in a structure that has many dead ends.

The reaction rate associated with a fractal depends both on the fractal dimension and on the way the reacting species move on the cluster. The combination of the two factors gives rise to another parameter called the spectral dimension. The quantity was introduced by Shlomo Alexander of the Hebrew University of Jerusalem and Raymond L. Orbach of the University of California at Los Angeles to describe diffusion and dynamics on a fractal. Although there is as yet no experimental evidence about the spectral dimension on nonequilibrium fractals, there is every reason to expect that it exists. In the future the very existence of the fractal geometry is likely to open still more frontiers of physics.

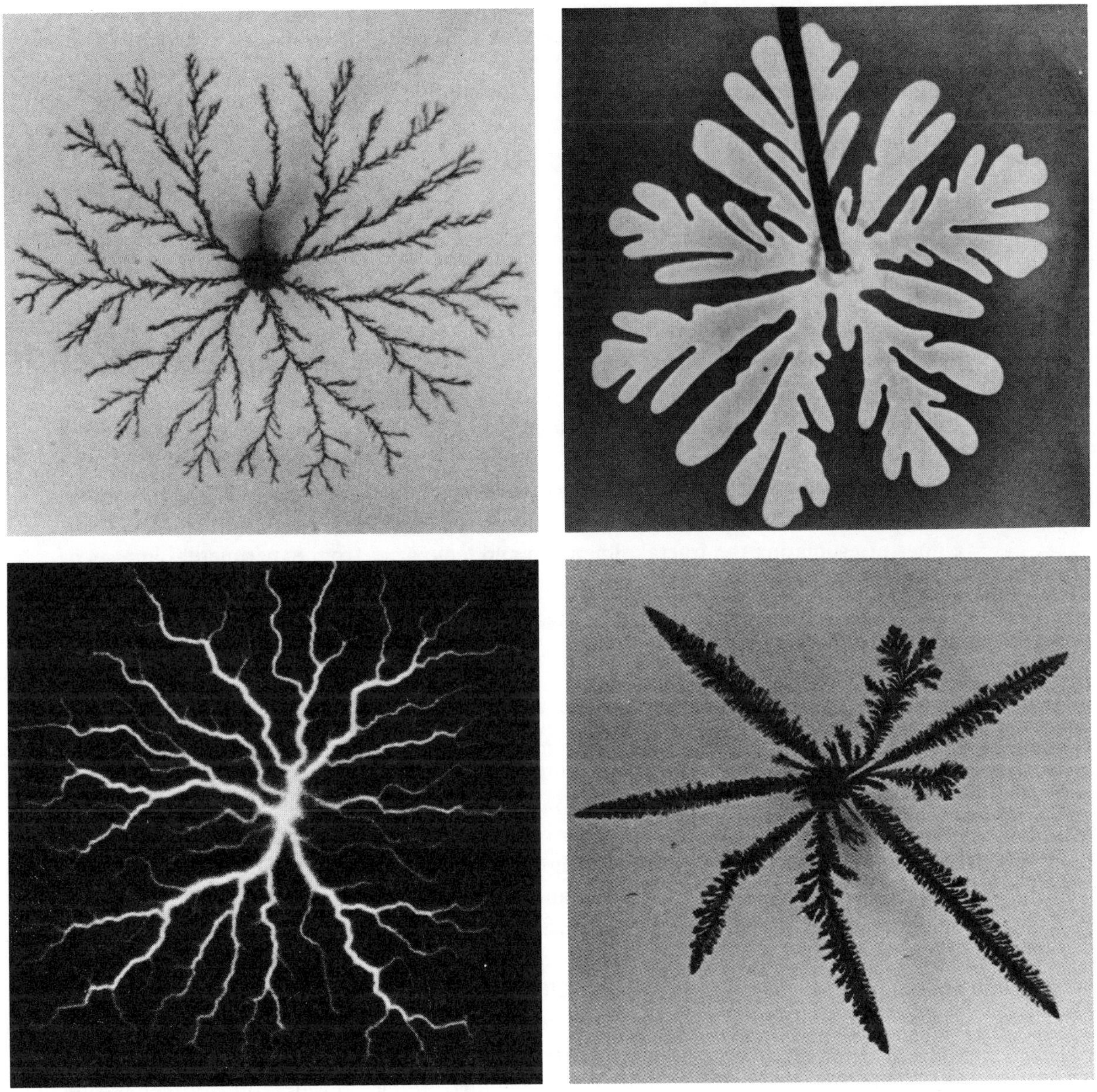

FRACTALS IN NATURE appear to grow by means of diffusion-limited aggregation. Shown here are a zinc deposit formed in an electrolytic cell (*top left*), a "viscous fingering" pattern of an air bubble in glycerine (*top right*) and an electrical-discharge pattern called a Lichtenberg figure (*bottom left*). The thick line that ends at the center of the bubble is an air tube. The zinc cluster at the bottom right shows what happens when the voltage in the electrolytic cell is increased: the growth pattern shifts from a fractal pattern to a dendritic, or snowflakelike, pattern. The zinc deposits were produced by Grier and the viscous-fingering pattern was produced by Eshel Ben-Jacob of the University of Michigan. The Lichtenberg figure is from L. Niemeyer and H. J. Wiesmann of Brown, Boveri & Company, Limited, in Switzerland, and Luciano Pietronero of the State University at Groningen.

Reprinted with permission from *Scaling Phenomena in Disordered Systems*, edited by R. Pynn and A. Skjeltorp, "Random Fractals: Characterization and Measurement," R.F. Voss, pp. 1–12, NATO ASI Series B **133**. Proceedings of the Geilo (Norway) Conference (April 1985), ©1986 Plenum Publishing, New York.

RANDOM FRACTALS: characterization and measurement

Richard F. Voss
IBM Thomas J. Watson Research Center
Yorktown Heights, NY 10598
and
Harvard University, Division of Applied Science
Cambridge, MA 01238

ABSTRACT

Mandelbrot's fractal geometry provides both a description and a mathematical model for many of the seemingly complex shapes found in nature. Such shapes often possess a remarkable invariance under changes of magnification. This statistical self-similarity may be characterized by a fractal dimension D, a number that agrees with our intuitive notion of dimension but need not be an integer. A brief mathematical characterization of random fractals is presented with emphasis on variations of Mandelbrot's fractional Brownian motion. The important concepts of fractal dimension and exact and statistical self-similarity and self-affinity will be reviewed. The various methods and difficulties of estimating the fractal dimension and lacunarity from experimental images or point sets are summarized.

random fractals: an introduction

Mandelbrot's fractal geometry [1] has revolutionized the application of geometrical constructs to the natural sciences. Fractals provide the proper mathematical framework for a treatment of the irregular, seemingly-complex shapes found in nature from the small scale structure of disordered systems and percolation clusters to coastlines, mountain ranges, clouds, and the distribution of stars in the night sky. Some of the building blocks of fractal geometry originated in the exactly self-similar mathematical "monsters" (such as the Koch curve and Seirpinski gasket) of the early 1900's. Although such exact deterministic constructs serve as useful tools in building understanding and intuition about scaling properties, the fractal shapes found in nature possess a statistical rather than exact self-similarity. The following sections present an expository summary of the major mathematical definitions and relations used in the characterization and measurement of random or statistical fractals as condensed from Mandelbrot[1]. A detailed discussion of the algorithmic considerations in generating or simulating such random fractals is found in [2].

fractional Brownian motion

One of the most useful mathematical models for the random fractals found in nature (such as mountainous terrain and clouds) has been the *fractional Brownian motion* (fBm) of Mandelbrot and Wallis[1,3]. It is an extension of the central concept of *Brownian motion* that has played an important role in both physics and mathematics. Sample *traces* of fBm are shown in Fig. 1. Almost all computer graphics fractal simulations[2] are based on an extension of fBm to higher dimensions such as the fractional Brownian landscape of Fig. 2. Fractional Brownian motion is also a good starting point for understanding anomalous diffusion and random walks on fractals.

A fractional Brownian motion, $V_H(t)$, is a single valued function of one variable, t (usually time). Its increments $V_H(t_2)$-$V_H(t_1)$ have a Gaussian distribution with variance

$$<|V_H(t_2) - V_H(t_1)|^2> \propto |t_2 - t_1|^{2H}, \tag{1}$$

where the brackets < and > denote averages over many samples of $V_H(t)$ and the parameter H has a value $0<H<1$. Such a function is both stationary and isotropic. Its mean square increments depend only on the time difference t_2-t_1 and all t's are statistically equivalent. The special value $H=1/2$ gives the familiar Brownian motion with $\Delta V^2 \propto \Delta t$.

As with the usual Brownian motion, although $V_H(t)$ is continuous, it is nowhere differentiable. Nevertheless, many constructs have been developed (and are relevant to the problem of light scattering from fractals) to give meaning to "derivative of fractional Brownian motion" as *fractional Gaussian noises*[1,3]. Such constructs are usually based on averages of $V_H(t)$ over decreasing scales. The derivative of normal Brownian motion, $H=1/2$, corresponds to uncorrelated *Gaussian white noise*, and Brownian motion is said to have *independent increments.* Formally, for any three times such that $t_1<t<t_2$, $V_H(t)$-$V_H(t_1)$ is statistically independent of $V_H(t_2)$-$V_H(t)$ for $H=1/2$. For $H>1/2$ there is a positive correlation both for the increments of $V_H(t)$ and its derivative fractional Gaussian noise. For $H<1/2$ the increments are negatively correlated. Such correlations extend to arbitrarily long time scales and have a large effect on the visual appearance of the fBm traces as shown in Fig. 1.

$V_H(t)$ shows a statistical scaling behavior. If the time scale t is changed by the factor r, then the increments ΔV_H change by a factor r^H. Formally,

$$<\Delta V_H(rt)^2> \propto r^{2H}<\Delta V_H(t)^2>. \tag{2}$$

Unlike statistically self-similar curves (such as the coastlines in Fig. 2), a $V_H(t)$ trace requires *different* scaling factors in the two coordinates (r for t but r^H for V_H) reflecting the special status of the t coordinate. Each t can correspond to only one value of V_H but any specific V_H may occur at multiple t's. Such non-uniform scaling is known as *self-affinity* rather than self-similarity.

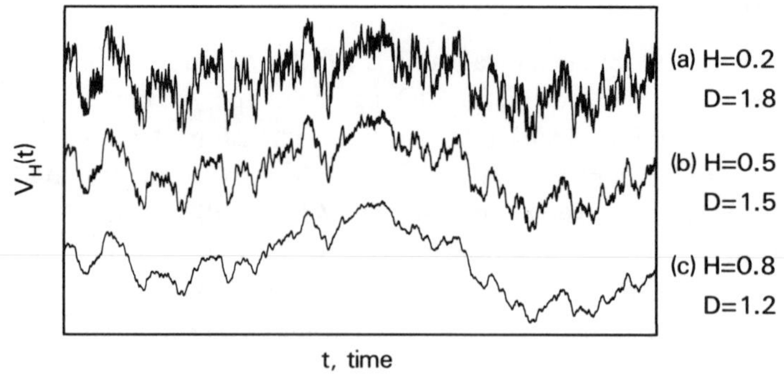

Figure 1. Sample traces of statistically self-affine fractional Brownian motion. (a) Increments are negatively correlated. (b) Normal Brownian motion with uncorrelated increments. (c) Increments are positively correlated.

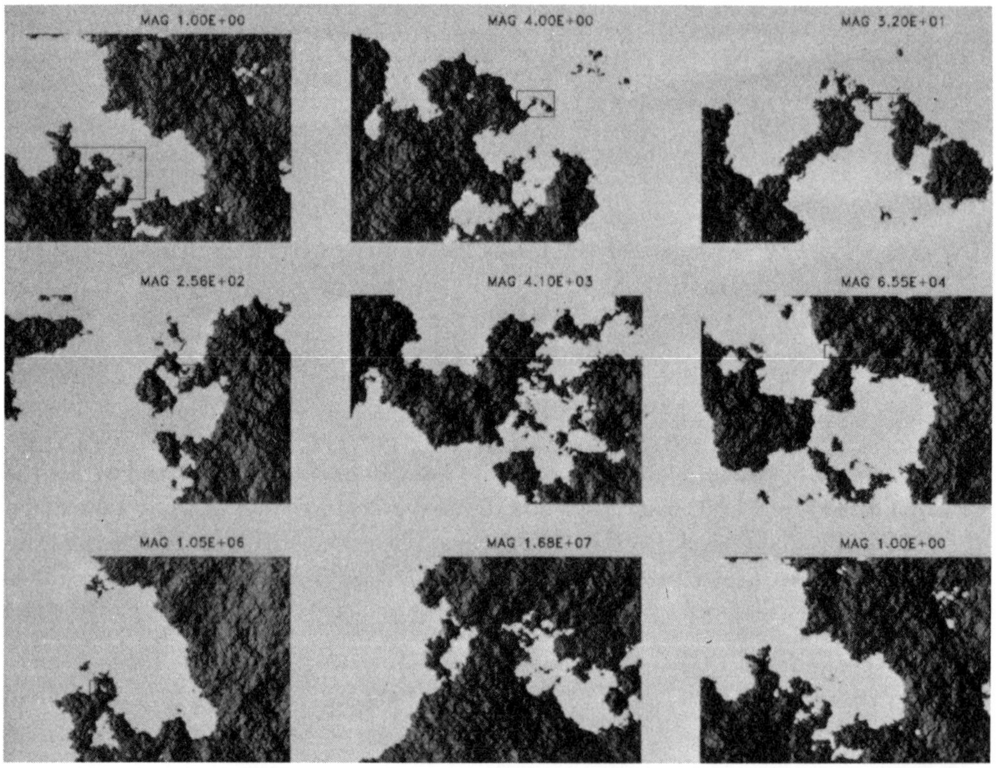

Figure 2. Computer realization of a statistically self-affine fractional Brownian landscape with H=0.8 D=2.2 at various magnifications. The altitude variations along any straight line path give a fBm trace such as Fig. 1(c). Successive frames show magnified view of the black outine from the previous frame and demonstrate that the landscape's coastline is statistically self-similar.

self-similar vs self-affine fractals.

The distinction between similarity and affinity is important. By way of summary, a *self-similar* object is composed of N copies of itself (with possible translations and rotations) each of which is scaled down by the ratio r in all E coordinates from the whole. More formally, consider a set S of points at positions $\vec{x} = (x_1, \ldots, x_E)$ in Euclidean space of dimension E. Under a *similarity* transform with real scaling ratio $0<r<1$, the set S becomes rS with points at $r\vec{x} = (rx_1, \ldots, rx_E)$. A bounded set S is *self-similar* when S is the union of N distinct (non-overlapping) subsets each of which is congruent to rS. *Congruent* means identical under translations and rotations. The fractal or *similarity dimension* of S is then given by

$$1 = Nr^D \quad \text{or} \quad D = \frac{\log N}{\log 1/r}. \tag{3}$$

This relation leads to several important methods of *estimating* D for a given set S.

For topologically one-dimensional fractal "curves", the apparent "length", varies with the measuring ruler size. If the entire self-similar curve is of maximum size L_{max}, then, on a smaller scale $L = rL_{max}$, the curve consists of $N = 1/r^D$ segments of length L. Thus, for $D \geq 1$

$$\text{LENGTH} = L \times N = L \times (L_{max}/L)^D \propto 1/L^{D-1}. \tag{4}$$

The fractal dimension D also characterizes the covering of the set S by E-dimensional "boxes" of linear size L. If the entire S is contained within one box of size L_{max}, then each of the $N=1/r^D$ subsets will fall within one box of size $L = rL_{max}$. Thus, the number of boxes of size L, $N_{box}(L)$, needed to cover S is given by

$$N_{box}(L) = (L_{max}/L)^D \quad \text{or} \quad N_{box}(L) \propto 1/L^D. \tag{5}$$

This definition of *box dimension* is one of the most useful methods for estimating the fractal dimension of a given set. The box dimension can be conveniently estimated by dividing the E-dimensional Euclidean space containing the set into a grid of boxes of size L^E and counting the number of such boxes $N_{box}(L)$ that are non-empty. It is useful to examine as large a range of L as possible and to average over various origins for the boxes.

One can also estimate the "volume" or "mass" of the set S by a covering with boxes of linear size L. If one considers only distances of order L *about a given point in S*, one finds a single box of size L with E-dimensional volume L^E. If the distance scale about the same point is increased to $L_{max}=L/r$, one now finds a total of $N=1/r^D=(L_{max}/L)^D$ boxes of mass L^E covering the set. Thus, the mass within a distance L_{max} of some point in S, $M(L_{max}) = N \times M(L) = M(L) \times (L_{max}/L)^D$ or

$$M(L) \propto L^D. \tag{6}$$

The fractal dimension D, thus, also corresponds to the commonly used *mass dimension* in physics. Mass dimension also has a strong connection with intuitive notions of dimension. The amount of material within a distance L of a point in a one-dimensional object increases as L^1. For an E-dimensional object it varies as L^E. The mass dimension is another extremely useful method for estimating the fractal dimension of a given object.

The set S is also *self-similar* if each of the N subsets is scaled down from the whole by a different similarity ratio r_n. In this case, D is given implicitly by

$$1 = \sum_{n=1}^{N} r_n^D , \tag{7}$$

which reduces to the familiar result in Eq. (3) when all of the r_n are equal.

The set S is *statistically self-similar* if it is composed of N distinct subsets each of which is scaled down by the ratio r from the original and is identical in all statistical respects to rS. The similarity dimension is again given by Eq. (3). In practice, it is impossible to verify that all moments of the distributions are identical, and claims of statistical self-similarity are usually based on only a few moments. Moreover, a sample of a random set (such as a coastline) is often statistically self-similar for all scaling ratios r. Its fractal dimension is usually estimated from the dependence of box coverings $N_{box}(L)$ or mass M(L) on varying L as in Eqs. (5) and (6).

Under an *affine* transform, on the other hand, each of the E coordinates of $\vec{x}$ may be scaled by a different ratio $(r_1, \ldots, r_E)$. Thus, the set S is transformed to r(S) with points at $r(\vec{x}) = (r_1x_1, \ldots, r_Ex_E)$. A bounded set S is *self-affine* when S is the union of N distinct (non-overlapping) subsets each of which is congruent to r(S). Similarly, S is *statistically self-affine* when S is the union of N distinct subsets each of which is congruent *in distribution* to r(S). The fractal dimension D, however, is not as easily defined as with self-similarity.

the relation of D to H for self-affine fractional Brownian motion

The assignment of a fractal dimension D to a self-affine set can be illustrated with a trace of fractional Brownian motion $V_H(t)$ from above. Consider, for convenience, a trace of $V_H(t)$ covering a time span $\Delta t=1$ and a vertical range $\Delta V_H=1$. $V_H(t)$ is statistically self-affine when t is scaled by r and V_H is scaled by r^H. Suppose the time span is divided into N equal intervals each with $\Delta t=1/N$. Each of these intervals will contain one portion of $V_H(t)$ with vertical range $\Delta V_H = \Delta t^H = 1/N^H$. Since $0<H<1$ each of these new sections will have a large vertical to horizontal size ratio and the occupied portion of each interval will be covered by $\Delta V_H/\Delta t = (1/N^H)/(1/N) = N/N^H$ square boxes of linear scale $L=1/N$. In terms of box dimension, as t is scaled down by a ratio $r=1/N$ the number of square boxes covering the trace goes from 1 to N(L) = number of intervals × boxes per interval or

$$N(L) = N \times N/N^H = N^{2-H} = 1/L^{2-H}. \tag{8}$$

Thus, by comparison with Eq. (5),

$$D = 2 - H \quad \text{for a trace of } V_H(t). \tag{9}$$

Consequently, the trace of normal Brownian motion has D = 1.5.

It is important to note that the association of a similarity dimension D with a self-affine fractal such as fBm is implicitly fixing a scaling between the (otherwise independent) coordinates. The result depends strongly on whether one is looking at scales large or small compared to this (artificially introduced) characteristic length. The difference is particularly clear when one attempts to estimate D for a trace of fBm from Eq. (4). As above, one can divide the t axis into N segments of size $\Delta t = 1/N$. For each of these segments the typical V variation will be $\Delta V = \Delta t^H$. The length *along* each segment is typically $\ell = (\Delta t^2 + \Delta V^2)^{1/2}$ and the total

$$\text{LENGTH} = N \times \ell \propto (1 + \Delta V^2/\Delta t^2)^{1/2} \propto (1 + 1/\Delta t^{2-2H})^{1/2}. \qquad (12)$$

On small scales with $\Delta t << 1$, the second term dominates and LENGTH $\propto 1/\Delta t^{1-H}$ so D = 2-H by comparison with Eq. (4) and in agreement with Eq. (9). On the other hand, on large scales with $\Delta t >> 1$, LENGTH is independent of Δt and D = 1.

The *zeroset* of fBm is the the intersection of the trace of $V_H(t)$ with the t axis, the set of all points such that $V_H(t) = 0$. The zeroset is a disconnected set of points with topological dimension zero and a fractal dimension $D_0 = D-1 = 1-H$ that is less than 1 but greater than 0. Although the trace of $V_H(t)$ is self-affine, its zeroset is self-similar.

trails of fBm

Consider a particle undergoing a fractional Brownian motion or random walk in E dimensions where each of the coordinates is tracing out an independent fBm in time. Over an interval Δt each coordinate will vary by typically $\Delta L = \Delta t^H$. If overlap can be neglected, the "mass" of the trail $M \propto \Delta t \propto L^{1/H}$. In comparison with Eq. (6), the trail of fBm has a fractal dimension

$$D = 1/H \qquad (13)$$

provided 1/H < E. Normal Brownian motion with H=0.5 has D=2. 1/H is known as the *latent* fractal dimension[4,5] of a *trail* of fBm. When 1/H > E overlap cannot be neglected and the actual D = E. When 1/H = E the trail is *critical* and most quantities will have important logarithmic corrections to the usual fractal power laws. Note that although the *trail* of a fBm in E dimensions is self-similar, each of the E coordinates has a self-affine *trace* vs time.

self-affinity in higher dimensions: Mandelbrot landscapes and clouds

The traces of fBm, particularly Fig. 1(c) with H=0.8, bear a striking resemblance to a mountainous horizon. The modelling of the irregular Earth's surface as a generalization of traces of fBm was first proposed by Mandelbrot. The single variable t can be replaced by coordinates x and y in the plane to give $V_H(x,y)$ as the surface altitude at position x,y as shown in Fig. 2. In this case, the altitude variations of a hiker following any straight line path at constant speed in the xy plane is a fractional Brownian motion. In analogy with Eq. (1),

$$< | V_H(x_2, y_2) - V_H(x_1, y_1) |^2 > \propto [(x_2 - x_1)^2 + (y_2 - y_1)^2]^H. \qquad (10)$$

Once again, the fractal dimension D must be greater than the topological dimension 2 of the surface. Here,

$$D = 3 - H \quad \text{for a fractal landscape, } V_H(x,y). \tag{11}$$

The intersection of a vertical plane with the surface $V_H(x,y)$ is a self-affine fBm trace with D=2-H, smaller by one than the value of Eq. (11). Similarly, the zeroset of $V_H(x,y)$, its intersection with a horizontal plane, also has a fractal dimension D_0=2-H. This intersection, which produces a family of (possibly disconnected) curves, forms the coastlines of the $V_H(x,y)$ landscape. Since the two coordinates x and y are, however, equivalent, the coastlines of $V_H(x,y)$ are self-similar, not self-affine. Figure 2 demonstrates how the coastline remains statistically invariant under changes of magnification.

This generalization of fBm can continue to still higher dimensions to produce, for example, a self-affine fractal temperature or density distribution $V_H(x,y,z)$. Here, the variations of an observer moving at constant speed along any straight line path in space generate a fBm and the fractal dimension

$$D = 4 - H \quad \text{for a fractal cloud } V_H(x,y,z). \tag{12}$$

The zeroset $V_H(x,y,z)$ = constant now gives a self-similar fractal with D_0=3-H.

To summarize, a statistically self-affine fractional Brownian function, V_H of $\vec{x} = (x_1, \ldots, x_E)$ in E Euclidean dimensions satisfies

$$<|V_H(\vec{x}_2) - V_H(\vec{x}_1)|^2> \propto |\vec{x}_2 - \vec{x}_1|^{2H}. \tag{13}$$

and has a fractal dimension

$$D = E + 1 - H. \tag{14}$$

The zerosets of $V_H(\vec{x})$ form a statistically self-similar fractal with dimension D_0 = E-H.

perimeter vs area scaling

The fractal dimension of the coastlines of $V_H(x,y)$ can also be estimated from the perimeter area scaling of the islands. Eq. (4) gives the LENGTH or perimeter of a fractal object of size L_{max} when "measured" with a ruler of size L. Consider a population of such islands with the same coastline D and all measured with the same ruler L. For a given island, perimeter $P \propto L_{max}^D$ from Eq. (4). Provided the coastline $D < 2$, each island will have a well-defined area $A \propto L_{max}^2$ and

$$P \propto A^{D/2}. \tag{15}$$

A study of P vs A scaling is a useful method of estimating a coastline D for a population of fractal objects.

spectral densities for fBm and the spectral exponent β

Random functions in time V(t) are often characterized[6,7] by their *spectral densities* $S_V(f)$. If V(t) is the input to a narrow bandpass filter at frequency f and bandwidth

Δf, then $S_V(f)$ is the mean square output $V(f)$ divided by Δf, $S_V(f) = |V(f)|^2/\Delta f$. $S_V(f)$ gives information about the time correlations of $V(t)$. When $S_V(f)$ increases steeply at low f, $V(t)$ varies more slowly. If one defines $V(f,T)$ as the Fourier transform of a specific sample of $V(t)$ for $0<t<T$,

$$V(f,T) = \frac{1}{T}\int_0^T V(t)e^{2\pi ift}dt, \text{ then } S_V(f) \propto T\,|V(f,T)|^2 \text{ as } T \rightarrow \infty. \qquad (16)$$

An alternate characterization of the time correlations of $V(t)$ is given by the *2 point autocorrelation function*

$$G_V(\tau) = <V(t)V(t + \tau)> - <V(t)>^2.$$

$G_V(\tau)$ provides a measure of how the fluctuations at two times separated by τ are related. $G_V(\tau)$ and $S_V(f)$ are not independent. In many cases they are related by the Wiener-Khintchine relation[6,7]

$$G_V(\tau) = \int_0^\infty S_V(f) \cos(2\pi f\tau)\, df. \qquad (17)$$

For a Gaussian white noise $S_V(f)$ = constant and $G_V(\tau) = \Delta V^2\delta(\tau)$ is completely uncorrelated. For certain simple power laws for $S_V(f)$, $G_V(\tau)$ can be calculated exactly. Thus, for

$$S_V(f) \propto 1/f^{\beta}, \text{ with } 0<\beta<1, \quad G_V(\tau) \propto \tau^{\beta-1}. \qquad (18)$$

Moreover, $G_V(\tau)$ is directly related to the mean square increments of fBm,

$$<|V(t + \tau) - V(t)|^2> = 2[<V^2> - G_V(\tau)]. \qquad (19)$$

Roughly speaking, $S_V(f) \propto 1/f^\beta$ corresponds to $G_V(\tau) \propto \tau^{1-\beta}$ and a fBm with $2H=\beta-1$ from Eqs. (1) and (19). Thus, the statistically self-affine fractional Brownian function, $V_H(\vec{x})$, with $\vec{x}$ in an E-dimensional Euclidean space, has a fractal dimension D and spectral density $S_V(f) \propto 1/f^\beta$, for the fluctuations along a straight line path in any direction in E-space with

$$D = E + 1 - H = E + \frac{3 - \beta}{2} \qquad (20)$$

This result agrees with other "extensions" of the concepts of *spectral density* and *Wiener-Khintchine relation* to *non-stationary* noises where some moments may be undefined. Moreover, it provides an extremely useful connection between D, H and β for finite simulations. For H in the range $0<H<1$, $E<D<E+1$, and $1<\beta<3$. The value $H \simeq 0.8$ is a good choice for many natural phenomena.

Although the formal definition of fBm restricts H to the range $0<H<1$, it is often useful to consider integration and an appropriate definition of "derivative" as extending the range of H. Thus, integration of a fBm produces a new fBm with H increased by 1, while "differentiation" reduces H by 1. When $H\rightarrow 1$, the derivative of fBm looks like a fBm with $H\rightarrow 0$. In terms of spectral density, if $V(t)$ has $S_V(f) \propto 1/f^\beta$ then its derivative dV/dt has spectral density $f^2/f^\beta = 1/f^{\beta-2}$. In terms of Eq. (20), differentiation decreases β by 2 and decreases H by 1.

measuring fractal dimensions: Mandelbrot measures

Numerical simulation or experimental image analysis often produces a geometrical object defined by a set S of points at positions $\vec{x} = (x_1, \ldots, x_E)$ in an E-dimensional Euclidean space. Lacking other information, all of the points may be assumed to be equivalent and all points are equally probable origins for analysis. The spatial arrangement of the points determines P(m,L). P(m,L) *is the probability that there are m points within an E-cube (or sphere) of size L centered about an arbitrary point in S.* P(m,L) is normalized

$$\sum_{m=1}^{N} P(m,L) = 1 \quad \text{for all } L \tag{21}$$

P(m,L) is directly related to other probability measures as used by Mandelbrot[1,8], Hentschel and Procaccia[9], and others. This particular definition of P(m,L) is, however, particularly efficient to implement on a computer.

The usual quantities of interest are derived from the moments of P(m,L). The *mass dimension*,

$$M(L) = \sum_{m=1}^{N} m\, P(m,L). \tag{22}$$

The number of boxes of size L needed to cover S,

$$N_{box}(L) = \sum_{m=1}^{N} \frac{1}{m} P(m,L). \tag{23}$$

The configurational entropy when space is divided into cubes of size L

$$S(L) = \sum_{m=1}^{N} \log m \;\; P(m,L). \tag{24}$$

For a fractal set $M(L) \propto L^D$, $N_{box}(L) \propto 1/L^D$, and $e^{S(L)} \propto L^D$.

In fact, one can define all moments

$$M^q(L) = \sum_{m=1}^{N} m^q\, P(m,L), \tag{25}$$

for $q \neq 0$ (the $q=0$ case is given by Eq.(24) above) and one can then estimate D from the logarithmic derivatives

$$D = \frac{1}{q} \left\langle \frac{\partial \log M_q(L)}{\partial \log L} \right\rangle \quad \text{for } q \neq 0, \tag{26}$$

and

$$D = \frac{1}{q} < \frac{\partial S(L)}{\partial \log L} > \quad \text{for } q = 0. \tag{27}$$

A double logarithmic plot of $M^q(L)$ vs L is an essential tool in verifying whether a fractal interpretation is valid for S. One expects a fractal to have power-law behavior of $M^q(L)$ over a wide range of L.

For a uniform fractal (fractal set) as the number of points examined, $N \to \infty$ the distribution is expected to take the scaling form $P(m,L) \to f(m/L^D)$ and all moments give the same D. For a non-uniform fractal (a fractal measure such as a Poincare map) the moments satisfy the relation[9] that $D_{q_1} \leq D_{q_2}$ for $q_1 < q_2$.

lacunarity

It is obvious from the above discussion that the fractal dimension D characterizes only part of the information in the distribution P(m,L). Different fractal sets may share the same D but have different appearances or *textures* corresponding to different P(m,L). As an initial step toward quantifying texture, Mandelbrot [1, chapter 34] has introduced the parameter *lacunarity*, Λ (*lacuna* is Latin for gap). Although the qualitative visual effect of changing lacunarity at fixed D is quite striking [1], to date there have been no quantitative measurements of the lacunarity of various random fractals and Mandelbrot [1] offers several alternative definitions. The most useful derives from the width of the distribution P(m,L) at fixed L. Given $M^q(L)$ as defined by Eq. (25),

$$\Lambda(L) = \frac{<M^2(L)> - <M(L)>^2}{<M(L)>^2}. \tag{28}$$

When P(m,L) takes the scaling form $f(m/L^D)$, Λ is just the relative mean square width of the distribution f, $\Lambda = <\Delta f^2>/<f>^2$.

acknowledgement

This work was made possible by the help, encouragement, and inspiration of Benoit Mandelbrot.

References

1. Mandelbrot, B.B. *The Fractal Geometry of Nature*, (Freeman, New York) 1982 and references therein. See also, *Fractals: Form, Chance, and Dimension,* W. H. Freeman and Co., San Francisco (1977).

2. Voss, R.F, "Random Fractal Forgeries:", *Proc. NATO A.S.I on Fundamental Algorithms in Computer Graphics,* Ilkley, Yorkshire, UK (1985).

3. Mandelbrot, B.B. and Wallis, J. W. "Fractional Brownian motions, fractional noises, and applications", *SIAM review, 10* (1968) 422-437.

4. Mandelbrot, B.B. "Fractals in Physics: Squig Clusters, Diffusions, Fractal Measures, and the Unicity of Fractal Dimensionality", *J. Stat. Phys. 34* (1984) 895-930.

5. Mandelbrot, B.B., Gefen, Y., Aharony, A., and Peyeiere, J. "Fractals, their transfer matrices and their eigen-dimensional sequences", *J. Phys. A 18* (1985) 335-354.

6. A good discussion is found in Reif, F. *Statistical and Thermal Physics,* McGraw-Hill Book Co., New York, (1965), Chapter 15, "Irreversible Processes and Fluctuations."

7. For example see: Freeman, J. J. *Principles of Noise,* John Wiley & Sons, Inc., New York, (1958), Chapter 1, "Fourier Series and Integrals." or Robinson, F.N.H. *Noise and Fluctuations,* Clarendon Press, Oxford, (1974).

8. Mandelbrot, B.B. "Intermittent turbulence in self-similar cascades: divergence of higher moments and dimension of the carrier", *J. Fluid Mech. 62* (1974) 331-358.

9. Hentschel, H.G.E. and Procaccia, I. "The infinite number of generalized dimensions of fractals and strange attractors", *Physica 8D* (1983) 435-444.

Reprinted with permission from *Physical Review Letters* **47**, T.A. Witten and L.M. Sander, "Diffusion-Limited Aggregation, a Kinetic Critical Phenomenon," pp. 1400–1403, ©1981 American Physical Society.

VOLUME 47, NUMBER 19 PHYSICAL REVIEW LETTERS 9 NOVEMBER 1981

Diffusion-Limited Aggregation, a Kinetic Critical Phenomenon

T. A. Witten, Jr.[a]
Groupe de Physique de la Matière Condensée, Collège de France, F-75231 Paris, France

and

L. M. Sander
Physics Department, University of Michigan, Ann Arbor, Michigan 48109
(Received 31 August 1981)

A model for random aggregates is studied by computer simulation. The model is applicable to a metal-particle aggregation process whose correlations have been measured previously. Density correlations within the model aggregates fall off with distance with a fractional power law, like those of the metal aggregates. The radius of gyration of the model aggregates has power-law behavior. The model is a limit of a model of dendritic growth.

PACS numbers: 68.70.+w, 05.70.Jk, 64.60.Cn, 82.70.Rr

Gold black, Cab-o-Sil™, and coagulated aerosols[1] are aggregates of solid particles distinguished by their wispy appearance. The size of these aggregates far exceeds the range of forces holding them together[2]; it seems likely that the aggregation process can be understood without reference to the details of these forces. A class of aggregates were shown by Forrest and Witten[3] to have density correlations of a power-law form. We show here that the formation conditions of these objects suggest a model which we simulate on a computer; the model is a discrete version of dendritic growth.[4] The simulation produces model aggregates which also have power-law correlations. We compare the behavior of the model to other models which produce low-density object: the Eden growth model,[5] random animals,[6] self-avoiding walks,[6] and percolating clusters.[5,7]

The aggregates which were studied by Forrest and Witten[3,8] were formed when a metal vapor produced by heating a plated filament was quench condensed. Metal particles of average radius 40 Å condensed near the filament and streamed outward. The particles accumulated in a thin spherical shell (the "puff ball") of roughly a centimeter radius. In this region the particles are thought to coalesce. Then they drifted down to an electron microscope slide. The aggregates found on the slides contained on the order of 10^5 metal particles in a low-density mass. The particles appeared to coalesce *irreversibly* to form the aggregates, since these maintained their wispy shape even after drifting through the gas. We suppose that two particles in the puff ball stick together whenever their thermal motion brings them into contact.

The simplest model for the growth of such a cluster of particles is the Eden model,[5] a lattice model in which particles are added one at a time at random to sites adjacent to occupied sites. This process produces a relatively compact cluster whose density correlations are independent of distance in the limit of large size. The metal aggregates, by contrast, were reported[3] to have correlations which fall off as a fractional power of distance. Thus they resemble random walks, percolating clusters, and other density profiles associated with critical phenomena.[5-7] We present here a model for the observed aggregation which also exhibits fractional power-law behavior of the correlation function.

Our model is a variant of the Eden model whose initial state is a seed particle at the origin of a lattice. A second particle is added at some random site at large distance from the origin. This particle walks randomly until it visits a site adjacent to the seed. Then the walking particle becomes part of the cluster. Another particle is now introduced at a random distant point, and it walks randomly until it joins the cluster, and so forth. If a particle touches the boundaries of the lattice in its random walk it is removed and another introduced. A similar process was studied by Rosenstock and Marquardt.[9] The exposed ends of our clusters tend to grow more rapidly than other perimeter sites because perimeter sites near the center are "shadowed"; our aggregates should be less compact than the Eden clusters.

Figure 1 shows a 3600-particle aggregate on a square lattice. Aggregates on a triangular lattice have the same general appearance. One may obtain information about the particle distribution from the density correlation function. The density $\rho(\vec{r})$ is defined to be 1 for the occupied sites

VOLUME 47, NUMBER 19 PHYSICAL REVIEW LETTERS 9 NOVEMBER 1981

FIG. 1. Random aggregate of 3600 particles on a square lattice.

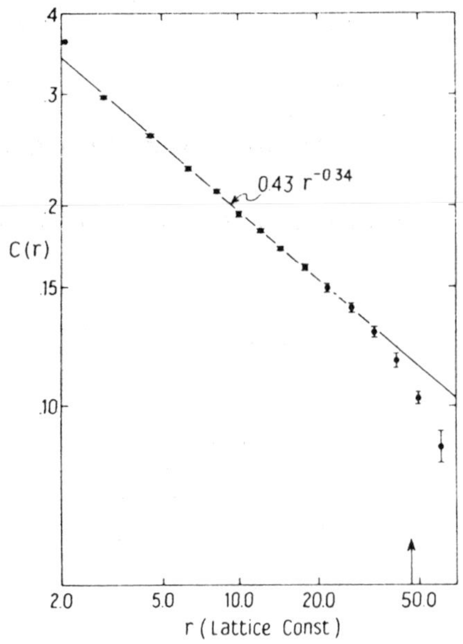

FIG. 2. The density correlation function averaged over six aggregates as a function of distance measured in lattice constants. The arrow marks the average radius of gyration. The solid line is a least-squares fit over the range $r=3$ to $r=27$. The error bars represent the spread of values among the six aggregates.

and 0 for the others. The correlation function in an N-particle aggregate,

$$C(r) \equiv N^{-1} \sum_{r'} \rho(r')\rho(r'+r), \tag{1}$$

is an approximation to the ensemble average correlation function $\langle \rho(\vec{r}')\rho(\vec{r}'+\vec{r})\rangle / \langle \rho(\vec{r}')\rangle$. This function is assumed to depend only on the distance r separating the two sites. Of course, this can only be true for r much less than the size of the aggregate. Figure 2 shows $C(r)$ averaged over directions and over six aggregates of roughly 3000 particles. It is clear from the fitted line that the data are consistent with a power law over distances from a few lattice spacings to the size of the cluster:

$$C(r) \sim r^{-0.343 \pm 0.004}. \tag{2}$$

The same exponent describes clusters on a triangular lattice; see Table I.

To check our computations we have measured[3] $C(r)$ for a certain Koch curve, a geometric construction which is a Hausdorff set with a Hausdorff dimension[10] D of $\ln 3/\ln 2 = 1.585$. One expects the density of points in a Hausdorff set to have correlations of the form

$$\langle \rho(\vec{r}')\rho(\vec{r}'+\vec{r})\rangle \sim r^{D-d} \equiv r^{-A}. \tag{3}$$

By definition, a Hausdorff set can be covered by a number of neighborhoods of radius a, $K(a)$, which varies as a^{-D}. The average number of points $N(a)$ within such a neighborhood must then vary as a^D. Assuming that the neighborhoods may be chosen with arbitrary centers, one may relate the behavior of $N(a)$ to that of the correlations:

$$a^D \sim N(a) = \int_0^a d^d r \langle \rho(\vec{r})\rangle_{\text{origin occupied}}$$
$$= \int_0^a d^d r \langle \rho(\vec{r})\rho(0)\rangle / \langle \rho(0)\rangle, \tag{4}$$

where d is the dimension of space. To yield this behavior the correlation function must obey Eq. (3).

To test the accuracy of our sampling procedure we generated images of the Koch curve and analyzed them with the same program used for the model aggregates. The resulting power-law fit confirms Eq. (3) as reported in Table I.

For the Koch curves and the model aggregates the measured $C(r)$ fall below the power-law line as r becomes comparable to the size of the aggregate. At short distances r the correlations of the model aggregates also differ from the power-law prediction. We believe this is a nonuniversal feature associated with the lattice.

The size R of a model aggregate should scale with the number of particles. Indeed, for neighborhoods of radius of order R, the contents $N(R)$

VOLUME 47, NUMBER 19 PHYSICAL REVIEW LETTERS 9 NOVEMBER 1981

TABLE I. Values of the correlation exponent A for diffusion-limited aggregation model and other systems.

	A Two dimensions	A Three dimensions
Diffusion-limited aggregation:		
Square lattice, average of six clusters of 2079–3609 particles in size (Fig. 2).	0.343 ± 0.004[a]	
Triangular lattice, average of three clusters of 1500–2997 particles.	0.327 ± 0.01[a]	
Radius of gyration, weighted average of A values inferred for six clusters of 999–3000 particles, Eq. (5).	0.299 ± 0.02[a]	
Koch curve with $A = 2D = 0.416$, measured average correlation function of seven curves, translated and rotated at random.	0.42	
Metal-particle aggregates, correlations of particle density from micrographs, Ref. 3.	0.32 ± 0.01[b]	1.32 ± 0.01[c]
Self-avoiding walk (flight), correlations of step density, Ref. 6.	0.667	1.33
Percolation, from radius of gyration of clusters at threshold, Eq. (5), Refs. 5 and 7.	~0.2	0.9
Random animals from radius of gyration, Eq. (5), Ref. 5.	0.46	1.18

[a]Error range indicates statistical error only.

[b]Here we interpret the measured two-dimensional density literally.

[c]Here we interpret the measured density profile as the two-dimensional projection of a three-dimensional profile and infer the A value for the three-dimensional density.

must approach N. Since $N(a) \sim a^D$ for $a \lesssim R$, we expect

$$N \sim R^D = R^{d-A}. \tag{5}$$

Measurements of the radii of gyration of our aggregates support this prediction, as indicated in Table I.

Figure 2 suggests[3] that our model aggregates are "critical" objects, having scale-independent correlations over an arbitrarily large range of distances. They are thus akin to self-avoiding walks, percolating clusters and order-parameter fluctuations near a phase transition. But in an important respect the model is different: The critical correlations here do not arise from an equilibrium ensemble but from an irreversible growth process. Indeed, our model is the discrete counterpart of the dendritic growth model of Langer and Muller-Krumbhaar.[4] In this continuum model a substance (e.g., one species in a two-component solution) diffuses from infinity to an absorbing surface. The density $u(\vec{r},t)$ of the substance is constant on the surface (neglecting surface-tension effects) and zero at infinity. The rate of growth of the surface at r is proportional to the flux density ∇u there.

In our model the probability that a perimeter site at $\vec{x}$ gains a particle at step n of the walk may also be expressed as a gradient of a quantity $u(\vec{r},n)$, which gives the probability (in the ensemble of all random walks of length n) that there is an nth step and that it is at $\vec{r}$. The probability $u(\vec{r},n)$ is proportional to the number of walks, $Z_n(\vec{r})$ arriving at $\vec{r}$ in n steps. This $Z_n(\vec{r})$ satisfies the relation

$$Z_{n+1}(\vec{r}) = \sum_{\alpha=1}^{c} Z_n(\vec{r}+\vec{\alpha}), \tag{6}$$

where $\vec{\alpha}$ is one of the "c" nearest-neighbor vectors. To obtain the number of walks which have not visited perimeter sites $\vec{x}$, one sets $Z_n(\vec{x}) = 0$

on the right side of Eq. (6), for these sites. This equation for Z_n implies a discrete diffusion equation for $u(\vec{r},n)=Z_n(\vec{r})/c^n$:

$$u(\vec{r},n+1)-u(\vec{r},n) = c^{-1}\sum_{\alpha=1}^{c}[u((\vec{r}+\vec{\alpha}),n)-u(\vec{r},n)], \quad (7)$$

with the boundary condition that $u=0$ on perimeter sites. The left side of Eq. (7) is a discrete derivative in n and the right-hand side is the discrete Laplacian operator. The probability $v_{n+1}(\vec{x})$ that the perimeter site at $\vec{x}$ grows in the $(n+1)$th step is

$$v_{n+1}(\vec{x})=\sum_{\alpha=1}^{c}u_n(\vec{x}+\vec{\alpha})/c = c^{-1}\sum_{\alpha=1}^{c}[u((\vec{x}+\vec{\alpha}),n)-u(\vec{x},n)]. \quad (8)$$

This is a discrete gradient at $\vec{x}$. Thus our model is a discrete version of the dendritic growth model described above.

In the study of this system by Langer and Muller-Krumbhaar,[4] they discovered that a smooth interface is unstable against wrinkling at all length scales; its growth depends on fluctuations in the diffusing field. Our findings describe the limit where these fluctuations are dominant.

The Monte Carlo study reported here may be extended to three or higher dimensions. We can study the effects of extending the size of the seed to simulate a finite tip radius of a dendrite. Realistic dendritic growth with surface tension can be simulated by making the sticking probability of particles smaller or larger for perimeter sites with fewer or more neighbors. Finally, the model can be used to predict the time-dependent growth rate of real random aggregates, as well as their mechanical and transport properties. The predicted features may appear in many processes, including gelation, condensation polymerization, and agglutination of biological molecules.

We are grateful to H. Gould, M. H. Cohen, J. J. Prentis, B. Roe, M. E. Sander, and R. Savit for valuable discussions. Supported by National Science Foundation Grants No. DMR-80-12867 and No. DMR-78-25012.

[(a)]Permanent address: Physics Department, University of Michigan, Ann Arbor, Mich. 48109.

[1]S. K. Friedlander, *Smoke, Dust, and Haze* (Wiley, New York, 1977).

[2]We neglect long-range electrostatic and magnetic forces.

[3]S. R. Forrest and T. A. Witten, Jr., J. Phys. A 12, L109 (1979).

[4]J. S. Langer and H. Muller-Krumbhaar, Acta Metall. 26, 1081, 1689 (1978); H. Muller-Krumbhaar and J. S. Langer, Acta Metall. 26, 1697 (1978).

[5]H. P. Peters, D. Stauffer, H. P. Hölters, and K. Loewenich, Z. Phys. B 34, 399 (1979).

[6]D. S. McKenzie, Phys. Rep. 27C, 37 (1976).

[7]H. E. Stanley, J. Phys. A 10, L211 (1977).

[8]S. Forrest, thesis, University of Michigan, 1979 (unpublished).

[9]H. B. Rosenstock and C. L. Marquardt, Phys. Rev. B 22, 5797 (1980). See also Z. Alexandrowicz, Phys. Lett. 80A, 284 (1980); P. Manneville and L. de Seze, to be published.

[10]B. Mandelbrot, *Fractals, Form, Chance, and Dimension* (Freeman, San Francisco, 1977).

Reprinted with permission from *Physical Review Letters* 51, M. Kolb, R. Botet, and R. Jullien, "Scaling of Kinetically Growing Clusters," pp. 1123–1126, ©1983 American Physical Society.

VOLUME 51, NUMBER 13 PHYSICAL REVIEW LETTERS 26 SEPTEMBER 1983

Scaling of Kinetically Growing Clusters

M. Kolb, R. Botet, and R. Jullien

Laboratoire de Physique des Solides, Université Paris-Sud, Centre d'Orsay, F-91405 Orsay, France

(Received 30 June 1983)

A model describing the process of cluster growth by aggregation of clusters is introduced and investigated with the Monte Carlo method. It is found that the clusters are scale invariant. The fractal dimension of this new class of kinetic critical behavior is $D = \nu^{-1} = 1.38 \pm 0.06$ in $d = 2$ dimensions as determined by the radii of gyration and the density correlations of the aggregates. The process is compared with other types of kinetic and equilibrium critical phenomena.

PACS numbers: 05.50.+q, 64.60.Cn, 82.70.Dd

The aggregation process, i.e., the mechanism of cluster formation from isolated single particles, is of interest in many areas of science such as physics (dendritic growth[1]), chemistry (flocculation of colloids,[2] formation of gels[3]), polymer physics (kinetics of polymerization[4]), medicine (growth of tumors[5]), and metereology and ecology (cloud formation, coagulation of smoke particles[6-8]).

The enormous progress made in the theory of critical phenomena has influenced the way one looks at the geometrical objects encountered in the above examples. This was recognized when a connection could be established between polymer statistics,[9] respectively percolation,[10] and critical phenomena that leads to a characterization of such systems in terms of scaling and universality.[11] Alternatively, the theory of fractals[12] describes the same phenomena on purely geometrical grounds, showing that it is not necessary to make reference to a statistical mechanical system governed by a Hamiltonian.

The field of critical behavior of clusters has been very active recently, particularly in the domain of growth models, whose irreversibility leads to scaling properties that are possibly different from those of equilibrium systems.[13]

In the present calculation, we establish scaling for a process encountered in a wide range of kinetic clustering phenomena, notably flocculation and coagulation. The model is a growth model that describes the clustering of clusters. One starts with an assembly of Brownian particles which stick together upon contact to form rigid clusters. The newly formed clusters diffuse along with the particles and continue to grow by aggregation when they meet other clusters or particles. The connection of this model with previously studied models is as follows: Single-particle diffusion-limited aggregation[8] has critical properties which are different from percolation and probably different from lattice animals.[14] The growth process under investigation resembles diffusion-limited aggregation in its kinetic mechanism. However, as we deal with many clusters, there are similarities to ordinary or kinetic percolation as used to describe gelation. As a consequence, one expects that the scaling properties are qualitatively different from both of the above situations. Our simulation indicates that, in two dimensions, the clusters are much more stringy than in all of these models. In the low-density regime, the model scales, suggesting that it may be self-similar, with a fractal dimension $D = 1.38 \pm 0.06$ as determined from the radii of gyration and the density correlations. As the clusters grow, the effective cluster density increases and eventually they cover the whole space.

The actual calculation is performed on a periodic square lattice with $V = L^2$ sites with initially N_0 randomly distributed particles. Clusters (particles connected by nearest-neighbor bonds) are rigid and move randomly one lattice spacing at a time. Clusters that touch (occupy nearest-neighbor sites) become part of a new, larger cluster. The clusters do not rotate. This idealized motion presumably does not affect the critical properties; thus a continuous movement of spherical particles forming rotating clusters would yield the same results.[14] According to our calculation, the type of kinetic motion does not affect the results. Most of the calculations were performed supposing that a cluster of N particles moves like a Brownian particle of mass N (probability to move $\propto 1/\sqrt{N}$), but the same exponents were obtained when we let all clusters move alike independent of their size or when heavy clusters hardly moved at all. Though the size distribution varies somewhat with the kinetic prescription, the qualitative aspects remain unchanged.

VOLUME 51, NUMBER 13 PHYSICAL REVIEW LETTERS 26 SEPTEMBER 1983

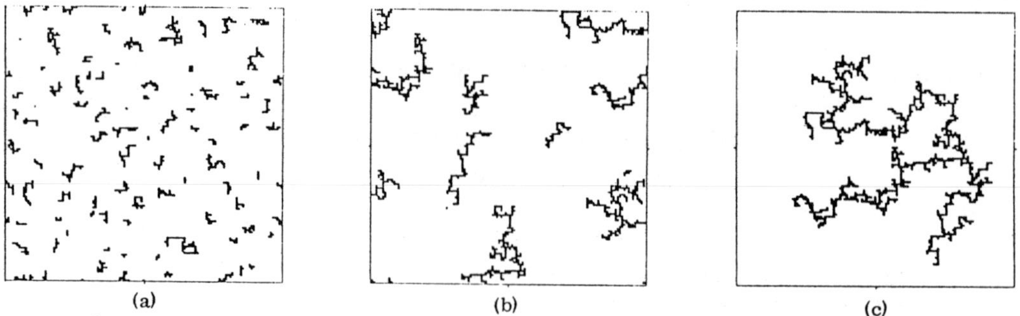

FIG. 1. Simulation of the growth of clusters. The three situations correspond to $N_c = 86$, 8, and 1 cluster, respectively. The system has a size $L = 128$ and initially $N_0 = 1024$ particles; the effective density grows from $\rho_0 = 0.06$ to $\rho_f \cong 1$. Note the chainlike appearance and the few branchings of the clusters.

The critical properties were extracted with use of two related measures characterizing the clusters. First, we determined the average size $R(N)$ of a cluster as a function of the number of particles in it. The radius of gyration behaves like

$$R \sim N^{\nu}, \quad N \sim R^{D}, \quad D = 1/\nu,$$

as $N \to \infty$. Secondly, we determined the density correlation function $C(\vec{r}) = \langle \rho(\vec{r}_0 + \vec{r}) \rangle$ at a point $\vec{r}_0$ of the cluster. It scales like

$$C \sim r^{-A}, \quad A + D = d, \quad 1 \ll r \ll R,$$

where d is the space dimension ($d = 2$). The averages are over all cluster configurations. In the simulation the average is taken over all the clusters with N particles for R and over all r_0 and all directions for $C(r)$.

In Fig. 1, the evolution of the clusters is shown for a system of size $L = 128$ with initially 1024 particles. As the number of clusters N_c decreases and the average number of particles per cluster $n = N_0/N_c$ increases, the size of a cluster is $V_c = R^d \sim n^{\nu d}$ and the effective density $\rho_{eff} = N_c(V_c/V) \sim \rho_0 n(d - D)/D$ grows for ramified ($D < d$) structures. No matter how low the initial concentration, the clusters eventually cover the whole space ($\rho_{eff} \sim 1$). The anticipated self-similarity under a change of scale then breaks down. In order to stay in the scaling region of interest, one has to lower the initial concentration with increasing system size L. In Fig. 2, the regions of different effective concentrations are indicated. The correlation function $C(r)$ distinguishes between an initial nonuniversal region and the following scaling region which terminates for low final concentration (all particles in one cluster) $\rho_f \ll 1$ at $r \sim R$ and for $\rho_f \lesssim 1$ crosses over to a more compact region. Our calculation determines the exponents A and ν and we find that this kinetic growth model is scale invariant, as shown in Figs. 3 and 4. The values $A = 0.61 \pm 0.05$ and $\nu = 0.73 \pm 0.04$ satisfy $d - A = D = 1/\nu$. As in other simulations of this kind we find that each simulation is quite smooth but that there are quite large statistical errors between different trials.[14]

There are other quantities which can be measured in the simulation. One, which explains why this model has very different critical properties than previously studied systems, is the size distribution of the clusters. It shows a broad and flat peak which allows clustering of clusters over

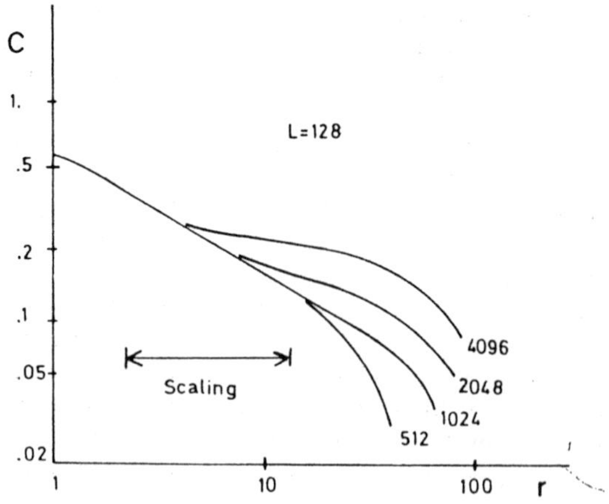

FIG. 2. Correlation function $C(r)$ (log-log plot) at different densities ρ_{eff}: crossover from dilute to compact region. The number of particles are $N_0 = 512$, 1024, 2048, and 4096, and the radii of gyration $R = 27$, 44, 51, and 55, respectively. While the radius of gyration limits the scaling region for low densities $\rho_f \ll 1$, the mutual hindrance of the clusters terminates it for $\rho_f \sim 1$.

VOLUME 51, NUMBER 13 PHYSICAL REVIEW LETTERS 26 SEPTEMBER 1983

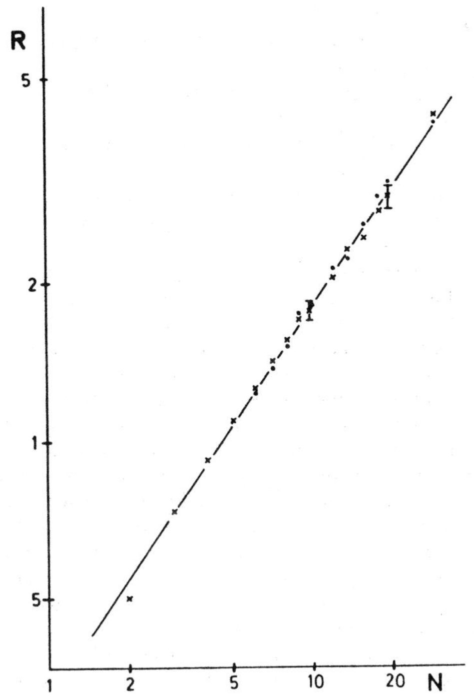

FIG. 3. Radius of gyration (log-log plot) in the scaling region. Average of ten samples with $L = 128$ and $N_0 = 1024$ with typical error bars (crosses) and two samples with $L = 450$ and $N_0 = 5000$ (dots). The straight line has slope 0.73.

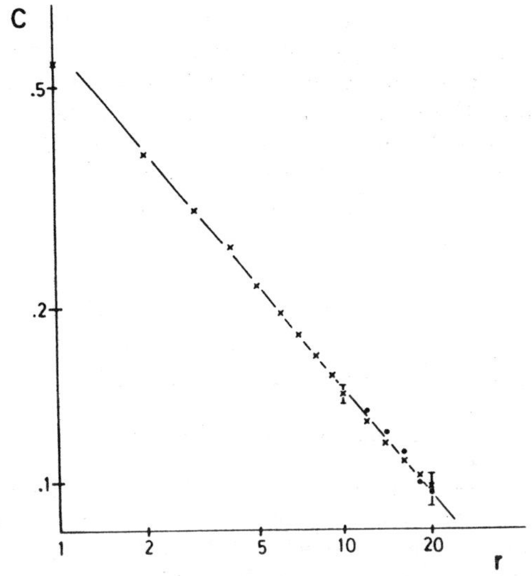

FIG. 4. Correlation function $C(r)$ for the same samples as in Fig. 3. The slope of the straight line is 0.61.

a wide range of sizes (provided $1 \ll N_c \ll N_0$), totally different from single-cluster aggregation.

Recently, Allain and Jouhier[15] observed clustering of clusters in a macroscopic experiment. At the percolation threshold[16] they find $\nu = 0.61$, intermediate between percolation and irreversible clustering. The experiment shows a slight tendency to restructure the clusters which may explain that the effective exponent $D_{\rm eff} > D$. Another aspect of their experiment is important: The motion of the clusters has both a random and a hydrodynamic element. Experimentally the influence of both parts cannot be separated. In our simulation we study the irreversible part alone.[17]

In Table I, the exponent $D = 1/\nu$ for kinetic and equilibrium critical phenomena are compared. The present model is far more ramified than lattice animals and single-particle aggregation. The linear appearance and the few branchings suggest a behavior not unlike linear polymers. In the light of the present calculation and the experiment of Allain and Jouhier,[15] the smoke-particle aggregation (Ref. 7) could be described by clustering of clusters with restructuring.[22]

The present study suggests various extensions. The simulation could be complemented with other approaches on the same model.[23] Also, it would be interesting to consider higher dimensions (preliminary results for $d = 3$ are analogous to $d = 2$: the clusters are much more ramified than other growth models, and D is close to the fractal dimension of linear polymers), time dependence,

TABLE I. Comparison of the fractal dimension $D = 1/\nu$ of the present model and of other kinetic and equilibrium critical phenomena. The estimate of D for the present model is based on the data of Figs. 3 and 4.

Phenomena	Exponent D ($d = 2$)
Present work	1.38 ± 0.06
Eden model[a]	2.0
Percolation[b]	1.80
Single-particle aggregation[c]	1.67
Lattice animals[d]	1.56
Linear polymers[e]	1.33
Smoke particles ($d = 3$)[f]	1.55–1.69
Experimental simulation[g]	1.64 ± 0.13

[a]Ref. 18.
[b]Ref. 19.
[c]Ref. 8.
[d]Ref. 20.
[e]Ref. 21.
[f]Ref. 7.
[g]Ref. 15.

the influence of hydrodynamic motion, and the crossover when partial restructuring is possible.

We would like to thank C. Allain for explaining in detail her experiment. This work was supported in part by the Fonds National Suisse. The Laboratoire de Physique des Solides is a Laboratoire associé au Centre National de Recherche Scientifique.

[1]J. S. Langer and H. Müller-Krumbhaar, Acta Metall. 26, 1681 (1978).

[2]D. H. Sutherland, J. Colloid Interface Sci. 25, 373 (1967).

[3]P. Manneville and L. de Seze, in *Numerical Methods in the Study of Critical Phenomena*, edited by I. Della Dora, J. Demangeot, and B. Lacolle (Springer-Verlag, Berlin, 1981); H. J. Herrman, D. Stauffer, and D. P. Landau, J. Phys. A 16, 1221 (1983).

[4]P. J. Flory, *Principles of Polymer Chemistry* (Cornell Univ. Press, Ithaca, 1953).

[5]T. Williams and R. Bjerknes, Nature (London) 236, 19 (1972).

[6]S. K. Friedlander, *Smoke, Dust and Haze: Fundamentals of Aerosol Behaviour* (Wiley, New York, 1977).

[7]S. R. Forrest and T. A. Witten, Jr., J. Phys. A 12, L109 (1979).

[8]T. A. Witten, Jr., and L. M. Sander, Phys. Rev. Lett. 47, 1400 (1981); H. Gould, F. Family, and H. E. Stanley, Phys. Rev. Lett. 50, 686 (1983); M. Muthukumar, Phys. Rev. Lett. 50, 839 (1983).

[9]P. G. de Gennes, Phys. Lett. 38A, 339 (1972).

[10]P. W. Kasteleyn and C. M. Fortuin, J. Phys. Soc. Jpn., Suppl. 26, 11 (1969).

[11]S.-K. Ma, *Modern Theory of Critical Phenomena* (Benjamin, Reading, Mass., 1976).

[12]B. M. Mandelbrot, *The Fractal Geometry of Nature* (Freeman, San Francisco, 1982).

[13]H. E. Stanley, F. Family, and H. Gould, to be published.

[14]P. Meakin, Phys. Rev. A 27, 604 (1983).

[15]C. Allain and B. Jouhier, J. Phys. (Paris), Lett. 44, L421 (1983).

[16]Allain and Jouhier originally tried to explain their experiment in terms of percolation (at a concentration when the largest cluster spans the system) and then found that the irreversible, kinetic clustering plays an important role (private communication).

[17]If no restructuring takes place the system is not related to percolation (the irreversibility guarantees a memory of the regime $\rho_{\rm eff} \ll 1$ even when later on the clusters percolate, Fig. 2) and simply becomes compact when $\rho_{\rm eff} \sim 1$. This is no longer true when it is allowed to restructure.

[18]M. Eden, in *Proceedings of the Fourth Berkeley Symposium on Mathematics, Statistics and Probability, Berkeley, 1960*, edited by Jerzy Neyman (Univ. of California Press, Berkeley, 1961), p. 233.

[19]H. E. Stanley, J. Phys. A 10, L211 (1977).

[20]B. Derrida and J. Vannimenus, J. Phys. (Paris), Lett. 41, L473 (1980).

[21]D. S. McKenzie, Phys. Rep. 27, 35 (1976).

[22]The clusters shown in Ref. 7 have an appearance that suggest a rather stringy structure, despite the projection from $d = 3$ to $d = 2$.

[23]A renormalization-group calculation is under way with encouraging results. It supports the notion that the relative size of the coalescing clusters determines the type of kinetic phenomena observed.

Reprinted with permission from *Physical Review Letters* **51**, P. Meakin, "Formation of Fractal Clusters and Networks by Irreversible Diffusion-Limited Aggregation," pp. 1119–1122, ©1983 American Physical Society.

VOLUME 51, NUMBER 13 PHYSICAL REVIEW LETTERS 26 SEPTEMBER 1983

Formation of Fractal Clusters and Networks by Irreversible Diffusion-Limited Aggregation

Paul Meakin

Central Research and Development Department, Experimental Station, E. I. du Pont de Nemours and Company, Wilmington, Delaware 19898

(Received 21 June 1983)

A model for diffusion-controlled aggregation in which growing clusters as well as individual particles are mobile has been investigated. Two versions of the model in which the cluster diffusion coefficient is either size independent or inversely proportional to number of particles (mass) give very similar results. In the limit of low concentration and large system size both models lead to structures with a fractal (Hausdorff) dimensionality of about 1.45–1.5 in two-dimensional lattice-based simulations.

PACS numbers: 05.40.+j, 64.60.Cn, 82.70.Dd

Witten and Sander[1] have recently developed a model for diffusion-limited aggregation which produces complex random dendritic structures. Numerical simulations using the Witten-Sander (WS) model indicate that these structures have remarkable scaling and universality properties.[1,2] In particular, the fractal[3] (Hausdorff[4]-Besicovitch) dimensionality D is distinctly smaller than the ordinary Euclidean dimensionality d (for $d = 2$, $D \approx \frac{5}{3}$).

These results have stimulated considerable theoretical[5-10] and experimental interest in diffusion-limited aggregation and other growth processes. The Witten-Sander model provides a basis for obtaining a better understanding of a variety of diffusion-limited processes such as random dendritic growth and the flocculation of colloidal systems. In the original Witten-Sander model all growth originates from a single immobile growth site and only one particle is allowed in the vicinity of the growing cluster at any time. These features are unrealistic for many real colloidal systems.

In this paper, I investigate a related model in which clusters of particles as well as single particles are allowed to "diffuse" and in which clusters of all sizes (including single particles) stick together on contact.

The two-dimensional simulations are carried out on a simple square lattice with periodic boundary conditions. At the start of the simulation, a fraction (ρ) of the lattice sites are picked at random and occupied (avoiding multiple occupancy). Sites (particles) at nearest-neighbor positions are considered to belong to the same cluster. Clusters (including single-particle clusters) are then picked at random and moved with a probability proportional to their "mobility" by one lattice spacing in one of four equally probable directions ($\pm x$, $\pm y$ also picked at random). If a cluster contacts other clusters (via nearest-neighbor occupancy), the contacting clusters are "merged" to form a single cluster. In this manner, the clusters grow larger and larger until only one large cluster remains. Two simple versions of the model have been investigated; in model A the cluster mobility is the same for all clusters and in model B the mobility is inversely proportional to the size (number of occupied sites) of the cluster. Other models which may be more relevant to specific physical systems could easily be implemented.

Figure 1 shows the final stage for several simulations carried out with a constant (mass-independent) mobility on 400×400 lattices. The densities are 0.0625 [Fig. 1(a)], 0.093 75 [Fig. 1(b)], 0.125 [Fig. 1(c)], and 0.156 particle per lattice site [Fig. 1(d)]. Density-density correlation functions obtained from these simulations are shown in Fig. 2 in the form of log-log plots. At short distances the plots of $\ln[C(r)]$ vs $\ln(r)$ are essentially linear with a slope which appears to approach a value of about $-\frac{1}{2}$ in the limit of low initial particle concentrations.

For a structure with a fractal[3] (Hausdorff[4]) dimensionality of D and a Euclidean dimensionality of d, we expect that the density-density correlation function $[C(r)]$ will have the form $C(r) \sim r^{-\alpha}$ where the exponent α is given by $\alpha = d - D$, i.e., α is the Hausdorff codimension. Many structures have a fractal geometry over a limited range of length scales with "cutoffs" at long and/or short lengths.[3] In this case, it appears that the lower cutoff length is a few lattice units and the upper cutoff length is determined by the average particle density $\langle\rho\rangle$. In other words there is a crossover from a fractal structure ($\alpha \approx -\frac{1}{2}$) on short length scales to a structure with uniform density

VOLUME 51, NUMBER 13 PHYSICAL REVIEW LETTERS 26 SEPTEMBER 1983

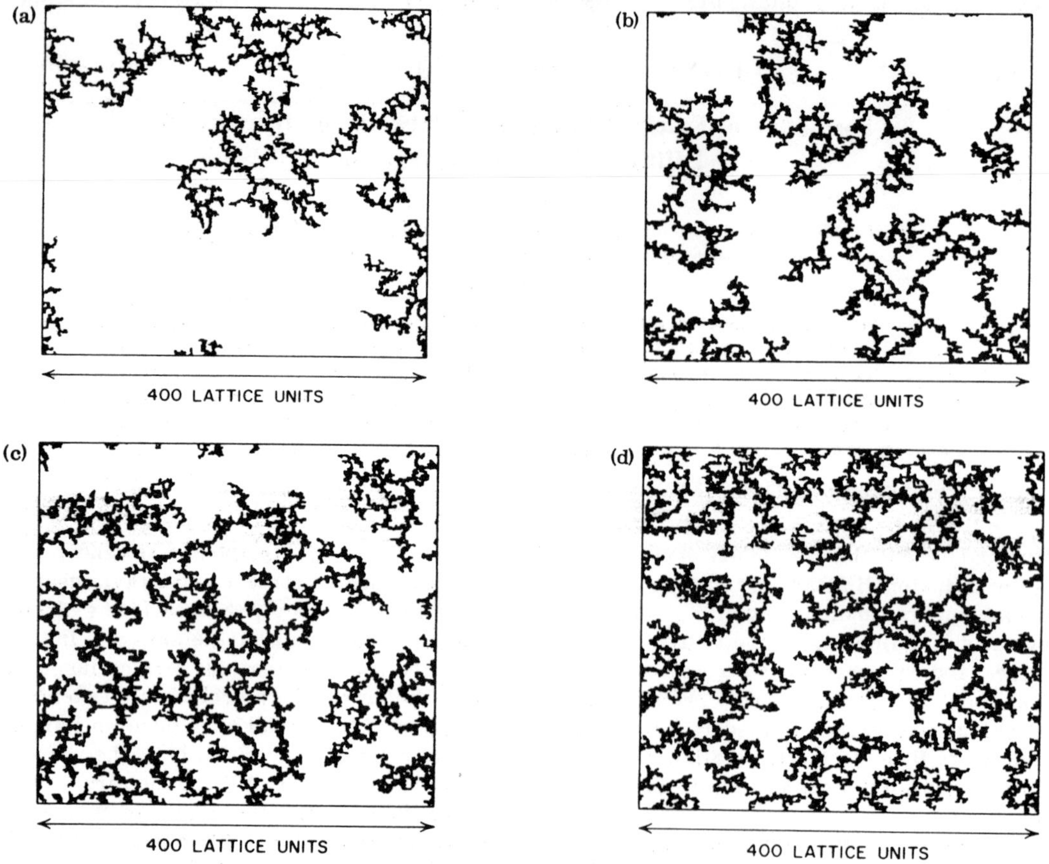

FIG. 1. The final stage of simulations carried out with a size-independent diffusion coefficient on a 400× 400 lattice. In (a) the cluster contains 10 000 particles (ρ = 0.0625 particle per lattice site); for (b), N = 15 000 (ρ = 0.093 75); for (c), N = 20 000 (ρ = 0.125); and for (d), N = 25 000 (ρ = 0.156 25).

on long length scales (α = 0) with a crossover length l given by

$$C_0 l^{-\alpha} \approx \langle\rho\rangle \text{ or } l \approx (\langle\rho\rangle/C_0)^{-1/\alpha},$$

where $C_0 \approx 1.0$. As $\langle\rho\rangle \to 0$, $l \to \infty$ and the resulting structures have a fractal geometry on essentially all length scales, if the present picture is correct.

In addition to the simulations used to obtain the results shown in Figs. 1 and 2, simulations have also been carried out at a lower concentration (5000 particles on a 400× 400 lattice or ρ = 0.031-25). From seven such simulations I find that $\alpha = 0.516 \pm 0.029$[11] ($1 \leq r \leq 25$ lattice units), $\alpha = 0.530 \pm 0.033$ ($5 \leq r \leq 25$), $\alpha = 0.491 \pm 0.009$ ($1 \leq r \leq 5$), and $\alpha = 0.524 \pm 0.028$ ($2 \leq r \leq 20$). A summary of these results is shown in Fig. 3.

Results very similar to those shown in Figs. 1 and 2 were obtained from simulations in which the cluster diffusion coefficient was inversely proportional to the cluster size (number of particles in the cluster). Figure 3 shows the density-density correlation function obtained from simulations carried out at a density of 0.031 25 particle per lattice site on a 400 × 400 lattice. The results shown in this figure represent the average of seven simulations. The Hausdorff codimension α is $\alpha = 0.519 \pm 0.014$ ($1 \leq r \leq 25$ lattice units), $\alpha = 0.519 \pm 0.037$ ($5 \leq r \leq 25$), $\alpha = 0.493 \pm 0.018$ ($1 \leq r \leq 5$), and $\alpha = 0.531 \pm 0.013$ ($2 \leq r \leq 20$). From these results I conclude that the limiting (low-density) Hausdorff dimensionality is 1.45–1.5 for both versions of the model. In Fig. 2 the dashed line with a slope of $-\frac{1}{3}$ represents the behavior expected for a very large object with a fractal dimensionality of (about) $\frac{5}{3}$ (such as a WS cluster). The dashed line with a slope of $-\frac{1}{2}$ represents the limiting (low-density,

VOLUME 51, NUMBER 13 PHYSICAL REVIEW LETTERS 26 SEPTEMBER 1983

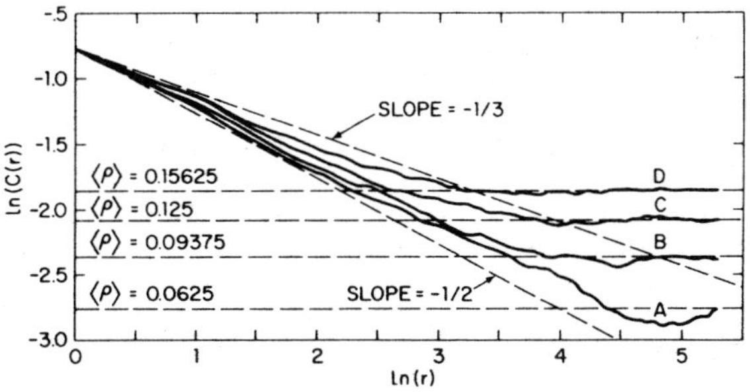

FIG. 2. The density-density correlation functions obtained from the structures shown in Fig. 1. Curve A is obtained from Fig. 1(a), curve B from Fig. 1(b), etc.

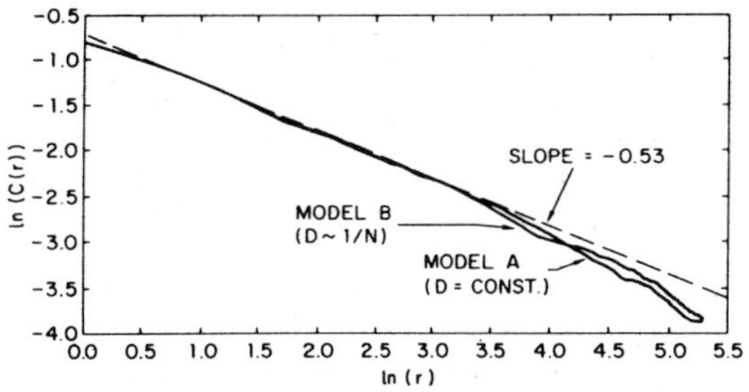

FIG. 3. Density-density correlation functions obtained from the final (single-cluster) stage of aggregation in simulations carried out at a density of 0.031 25 particle per lattice site. Curve A was obtained from model A (size-independent diffusion coefficient) and curve B from model B (diffusion coefficient proportional to 1/mass). Both curves represent the average of seven simulations carried out with 5 000 particles on 400×400 lattices. The dashed line has a slope of −0.53 and represents the correlation functions for a structure with a Hausdorff dimensionality of 1.47 on all length scales. The deviation of the correlation functions from the dashed line at large distances (r) is due to finite-size effects.

large-size) behavior suggested by our results. The fractal dimensionality has also been estimated from the dependence of the cluster radius of gyration R_g on cluster size N from the relationship $R_g \sim N^{1/D}$. Results obtained from fitting a straight line to the dependence of $\ln(R_g)$ on $\ln(N)$ for intermediate clusters formed during the course of the simulations described above indicate that $D \approx 1.4-1.5$ for both models A and B.

A few simulations have also been carried out with other similar models. If only the *smallest* cluster(s) are allowed to move, results very similar to those obtained with models A and B (above) were obtained ($D \approx 1.5$ in the small concentration limit). In this model, single particles are removed during the early stages of the simulation and the smallest clusters become doublets which in turn are rapidly removed leaving only larger clusters in the system. If only the *largest* cluster is allowed to move the model becomes equivalent to the Witten-Sander model of diffusion-limited aggregation in the limit $\rho \to 0$ (i.e., $D \approx \frac{5}{3}$ for $d=2$). If ρ is finite a crossover from $D \approx \frac{5}{3}$ on short length scales to $D \approx 2.0$ on long length scales is observed. This crossover effect is similar to that found with models A and B. Similar behavior has also been observed in computer simulations using a modified Witten-Sander model[12] and has been predicted theoretically by Nauenberg, Richter, and Sander[10] for large aggregates grown at a finite density of diffusing particles.

The fractal dimensionalities of the structures generated using these models are distinctly different from the dimensionalities of most other random structures in two dimensions such as percolation clusters at the percolation threshold ($D \approx 1.9$),[13] self-avoiding random walks ($D \approx 1.33$),[14,15] and Witten-Sander clusters ($D \approx 1.67$).[1] However, two-dimensional lattice animals do have a fractal dimensionality which is quite close (≈ 1.56).[16]

The models described in this paper seem to have the same sort of universality properties as the WS model[1,2] of diffusion-limited aggregates. The fractal dimensionality is insensitive to the sticking probability and the effective dimensionality obtained from a nonlattice version of the model[17] is equal to that obtained from the lattice models described above (within the accuracy of the simulations). I am presently in the process of carrying out very similar simulations using three-dimensional lattices and related nonlattice simulations in two and three dimensions. The model described in this paper includes translational but no rotational diffusion. A nonlattice model will allow both effects to be included. Sutherland and Goodarz-Nia[18] have investigated three-dimensional nonlattice models in which clusters aggregate via linear (rather than Brownian) trajectories. Their results may be interpreted in terms of an effective dimensionality of about 1.85. Using a two-dimensional version of the Sutherland–Goodarz-Nia model, I have generated fractal aggregates with a dimensionality similar to that associated with the present model.

The process(es) simulated in this paper are closely related to spinodal decomposition with very deep quenches. Under these conditions Heerman and Klein[19] have found evidence that nucleating droplets formed during Monte Carlo simulations of nucleation and growth in Ising systems with Glauber dynamics have noncompact (fractal) structures during the initial stages of growth.

The model described in this paper has been discussed with many colleagues during the past year or so. The author would like to thank them for their suggestions and encouragement. In particular helpful discussions with T. A. Witten are acknowledged.

[1]T. A. Witten and L. M. Sander, Phys. Rev. Lett. 47, 1400 (1981).
[2]P. Meakin, Phys. Rev. A 27, 604, 1495 (1983).
[3]B. B. Mandelbrot, *The Fractal Geometry of Nature* (Freeman, San Francisco, 1982).
[4]F. Hausdorff, Math. Ann. 79, 157 (1919).
[5]M. Muthukumar, Phys. Rev. Lett. 50, 839 (1983).
[6]H. Gould, H. E. Stanley, and F. Family, Phys. Rev. Lett. 50, 686 (1983).
[7]M. Nauenberg, to be published.
[8]R. Ball, M. Nauenberg, and T. A. Witten, to be published.
[9]T. A. Witten and L. M. Sander, Phys. Rev. B 27, 5686 (1983).
[10]M. Nauenberg, R. Richter, and L. M. Sander, to be published.
[11]The uncertainties given in this paper represent 95% confidence limits. Only statistical uncertainties are included. Errors due to finite-size effects, etc., may be significantly larger.
[12]T. A. Witten and P. Meakin, to be published.
[13]H. E. Stanley, J. Phys. A 10, 2211 (1977).
[14]D. S. McKenzie, Phys. Rep. 27, 35 (1976).
[15]S. Havlin and D. Ben-Avraham, Phys. Rev. A 26, 1728 (1982).
[16]H. E. Stanley, P. J. Reynolds, S. Redner, and F. Family, in *Real Space Renormalization*, edited by T. W. Burkhardt and J. M. J. van Leeuwon (Springer-Verlag, Berlin, 1982).
[17]P. Meakin and Z. R. Wasserman, to be published.
[18]D. N. Sutherland and I. Goodarz-Nia, Chem. Eng. Sci. 26, 2071 (1971).
[19]D. W. Heerman and W. Klein, Phys. Rev. Lett. 50, 1062 (1983).

Reprinted with permission from *Physical Review Letters* **57**, P. Dimon, S.K. Sinha, D.A. Weitz, C.R. Safinya, G.S. Smith, W.A. Varady, and H.M. Lindsay, "Structure of Aggregated Gold Colloids," p. 595–598, ©1986 American Physical Society.

VOLUME 57, NUMBER 5 PHYSICAL REVIEW LETTERS 4 AUGUST 1986

Structure of Aggregated Gold Colloids

P. Dimon, S. K. Sinha, D. A. Weitz, C. R. Safinya, G. S. Smith,[a] W. A. Varady, and H. M. Lindsay
Exxon Research and Engineering Company, Annandale, New Jersey 08801
(Received 19 February 1986)

We report a high-resolution, small-angle x-ray study of aggregated gold colloids over the range 0.0003 to 0.08 Å^{-1}. We are able to fit our data with a simple model that correctly accounts for nonfractal short-range order with a crossover to long-range fractal correlations. This provides new information on the structure of real aggregates, and new insight into the aggregation processes which lead to their formation.

PACS numbers: 64.60.Cn, 05.40.+j, 61.10.Lx, 82.70.Dd

What is the structure of an aggregate? Several years ago, Witten and Sander[1] showed that aggregates can be statistically scale-invariant objects, physical realizations of mathematical fractals[2] which appear the same on all length scales. This property manifests itself in a power-law dependence of the density-density correlation function, i.e., $G(r) \sim r^{D-d}$, where d is the spatial dimension and D is the fractal dimension $(D < d)$. These results have been elucidated by computer simulations of several different models including the diffusion-limited aggregation (DLA) model,[1,3] the diffusion-limited cluster aggregation (DLCA) model,[4] and the reaction-limited cluster aggregation (RLCA) model.[5] Experimentally, aggregates of gold colloids[6–8] and silica particles[9–11] have been studied with light, neutron, and low-resolution x-ray scattering. The emphasis of most studies to date has been the fractal nature of the aggregates. However, in reality, objects can be scale invariant only over a limited range, and an understanding of the bounds of the fractal regime is crucial in the development of a complete picture, not only of aggregates, but of fractal objects in general. In particular, as we shall see, the short-range order, or correlations of the nearest neighbors, cannot be treated as fractal. However, these correlations play an important role in determining the physical properties of the ultimate structure and, in addition, offer important clues about the interparticle interactions involved in the aggregation process itself. In this paper, we report the results of a high-resolution x-ray scattering experiment using colloidal gold aggregates, and present data that extend over both the conventional light and small-angle x-ray and neutron scattering regimes. This study, with momentum transfers (length scales) continuously ranging from 0.0003 Å^{-1} ($\sim$3000 Å) to 0.08 Å^{-1} ($\sim$10 Å) and on samples comprising over 10^6 clusters, provides the most definitive probe of the fractal structure of aggregates. The excellent statistics and the wide range of these data allow us to probe both the fractal regime of the aggregate structure and the crossover regime where the nonfractal short-range order dominates. A simple model is developed to account quantitatively for these data, and the short-range correlations thus determined are related to the kinetic growth processes that form the aggregate.

Three samples of aggregated gold colloids comprised of $\sim$75-Å-radius, polycrystalline gold balls[6] were studied. The colloids were aggregated by reduction of the charge on the surface of the particles causing them to stick upon diffusion-induced collision. In one case, the charge was completely removed so that the diffusion of the clusters was the rate-limiting process, resulting in DLCA. In the second case, the charge was only slightly reduced, so that the actual chemical bond formation was the rate-limiting process, resulting in RLCA. The scattering was done from clusters large enough to precipitate, thus increasing the volume fraction of gold from its initial value of 10^{-6} to roughly 10^{-3}. The precipitate solution was placed between two 0.127-mm Kapton sheets. The experiment was done on beamline VI-2 at the Stanford Synchrotron Radiation Laboratory with 1.54-Å x rays. We used a double-bounce Si(111) monochromator and a triple-bounce Si(111) analyzer in the nondispersive configuration to obtain a sharp Gaussian in-plane resolution function with rms width 6×10^{-5} Å^{-1}. The out-of-plane resolution function was determined by a set of extremely narrow slits, yielding a Gaussian out-of-plane resolution function of rms width 3.4×10^{-4} Å^{-1}. The sharpness of the resolution function enabled us to reach momentum transfers of 0.0003 Å^{-1} before we were swamped by the main beam.

Figure 1 shows the background-subtracted data for the RLCA sample, corrected for the transmission factor. At small Q, the data clearly show power-law behavior reflecting the long-range fractal correlations, while at large Q, the behavior is dominated by the form factor of the individual gold balls comprising the aggregate. However, a model which includes only a fractal $S(Q)$ modified by the form factor of a single gold ball, as was used to account for previous neutron- and light-scattering data from gold aggregates,[8] cannot describe the present, more extensive data. To account for these data we must include short-range nonfractal correlations. Physically, we expect that $G(r) \sim r^{D-3}$ should only be valid for r greater than a few ball diam-

VOLUME 57, NUMBER 5 PHYSICAL REVIEW LETTERS 4 AUGUST 1986

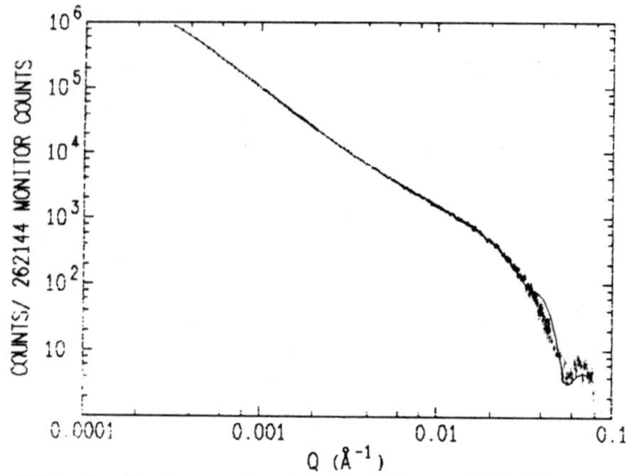

FIG. 1. Background-subtracted data of RLCA sample. The solid line is a fit by our model (see text). The range in Q probed corresponds to length scales of ~ 3000 Å (0.0003 Å^{-1}) to ~ 10 Å (0.08 Å^{-1}). The background level was $\sim 1\%$ of the total signal.

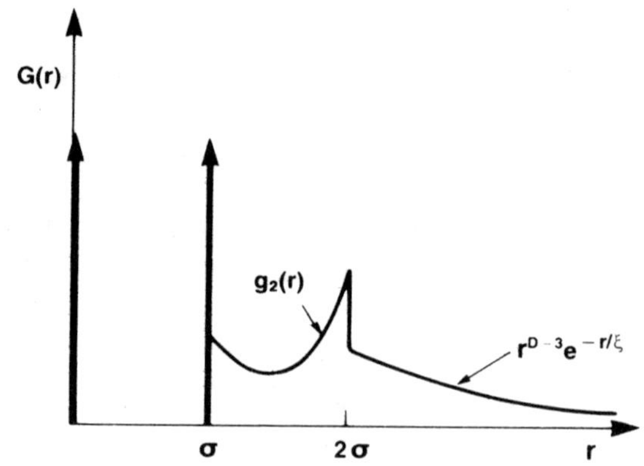

FIG. 2. Correlation function $G(r)$ used in our model [see Eq. (1)].

eters. For shorter distances, we can model the aggregate as being comprised of hard spheres in close contact. If we assume an isotropic aggregate, the correlation function can be expressed in the following form (see Fig. 2):

$$G(r)=\begin{cases}\delta(\mathbf{r})+(z_1/4\pi\sigma^2)\delta(r-\sigma) & (r\leq\sigma),\\ z_2g_2(r) & (\sigma<r<2\sigma),\\ A_fBr^{D-3}e^{-r/\xi} & (r>2\sigma).\end{cases} \quad (1)$$

The δ function at the origin is due to self-correlations. The δ function at $r=\sigma$, where σ is a ball diameter, reflects the presence of nearest neighbors in hard contact with the ball at the origin. Thus, z_1 is the coordination number of this first shell. The second line of Eq. (1) represents the second shell, whose presence was inferred from computer simulations of both random-packed hard spheres[12] and DLCA aggregates.[13] $g_2(r)$ is a second-order polynomial fit to the hard-sphere simulation data with $4\pi\int_\sigma^{2\sigma} g_2(r)r^2dr=1$ so that z_2 is the second-shell coordination number. We found that including this term made the fits somewhat better but was not essential. The last line of Eq. (1) describes the crossover to long-range fractal correlations. (A correlation length ξ has been included to account for overlap among clusters,[10] but because of their size, it was found to be effectively infinite.) B is a constant defined so that $G(r)$ is continuous at $r=2\sigma$ when $A_f=1$, i.e., $z_2g_2(2\sigma)=B(2\sigma)^{D-3}e^{-2\sigma/\xi}$. A_f therefore measures the amplitude of the fractal term relative to the second-shell term at $r=2\sigma$. There are thus four adjustable parameters in the model: D, z_1, z_2, and A_f.

The structure factor for the aggregate is then given by the Fourier transform of $G(r)$:

$$S_A(Q)=1+z_1\sin Q\sigma/Q\sigma+(4\pi z_2/Q)\int_\sigma^{2\sigma} rg_2(r)\sin Qr\,dr+(4\pi A_fB/Q)\int_{2\sigma}^{\infty} r^{D-2}e^{-r/\xi}\sin Qr\,dr. \quad (2)$$

The second integral in Eq. (2) can be expressed as an incomplete gamma function and evaluated numerically by series expansions. The actual measured intensity is given by

$$I(\mathbf{Q})=\int |f(Q'r_0)|^2S_A(Q')R(\mathbf{Q}-\mathbf{Q}')d^3Q', \quad (3)$$

where $f(Qr_0)$ is the form factor of a gold ball of radius r_0 and $R(\mathbf{Q}-\mathbf{Q}')$ is the resolution function discussed earlier. For small-angle scattering geometry, we can take

$$R(\mathbf{Q}-\mathbf{Q}')\sim\exp[-(Q_x-Q_x')^2/2\sigma_x^2]\exp(-Q_z'^2/2\sigma_z^2)\delta(Q_y'),$$

where Q_x and Q_y are in the scattering plane, and the scan is done along Q_x. As stated earlier, $\sigma_x=6\times10^{-5}$ Å^{-1} and $\sigma_z=3.4\times10^{-4}$ Å^{-1}. Proper inclusion of the resolution function is essential as it will always result in the fitted fractal dimension being slightly larger ($\sim 10\%$ in our experiments) than if it were measured directly from the data.

It is also necessary to consider the effects of polydispersity in the analysis. So far it has been assumed that the gold balls are uniform in size. If this were true, the first minimum in the form factor at ~ 0.06 Å^{-1} would be identically zero resulting in a sharp dip where we see only a shallow one. In a polydisperse system we have

$$S(\mathbf{Q})=\frac{1}{N}\sum_i|f_i(\mathbf{Q})|^2+\frac{1}{N}\sum_{i\neq j}f_i(\mathbf{Q})f_j^*(\mathbf{Q})\exp[i\mathbf{Q}\cdot(\mathbf{r}_i-\mathbf{r}_j)],$$

or, replacing the sums by averages, we can substitute

$$|f(Qr_0)|^2 S_A(Q) \rightarrow \langle |f(Qr_0)|^2\rangle + |\langle f(Qr_0)\rangle|^2 [S_A(Q)-1]$$

in Eq. (3), where $\langle\ \rangle$ signifies an average over the distribution of ball sizes. If the distribution is Gaussian, then for any function $h(r_0)$,

$$\langle h(r_0)\rangle = \frac{1}{(2\pi)^{1/2}\delta_P}\int_0^\infty h(r_0)\exp[-(r_0-\bar{r})^2/2\delta_P^2]dr_0,$$

where $\bar{r}$ is the mean radius, and δ_P is the rms width. The normalization constant is based on the assumption that the unphysical range $(-\infty,0)$ contributes negligibly in the integral if δ_P is small compared with $\bar{r}$. We found empirically that $\bar{r}=79$ Å and $\delta_P=0.1r$ reproduced the position and size of the dip in the RLCA data. This is actually consistent with previous transmission-electron-microscopy measurements which found $\bar{r}\sim 75$ Å and rms deviations from 8% to 13%.[14] Different values for the polydispersity did not affect the fits greatly. Polydispersity will also smear the form of $S_A(Q)$ in Eq. (2), but these effects were investigated and found to be unimportant.

The result of the fitting of Eq. (3) to the RLCA data is shown in Fig. 1. We find that $D=2.20$, a number somewhat higher than previous results.[7] This number is essentially model independent (not including resolution effects), being determined primarily by the data at small Q where the statistics are superior and short-range effects are unimportant. The first-shell coordination number is $z_1=2.33$, which, considering this was a free parameter, is not unreasonable for such a tenuous object as an aggregate. (In his simulations of the DLA model, Meakin[3] found $z_1=2.25$.) Also, $z_2=3.95$, which is larger than z_1, as it should be. The parameter A_f is 0.46, which is consistent with computer simulations of random packed hard spheres[12] and aggregates.[13] The small shoulder in the fit at $Q\sim 0.04$ Å^{-1} arises from the δ function in $G(r)$ at $r=\sigma$ and probably reflects our imperfect knowledge of $G(r)$.

Our model is not as successful in accounting for the complete set of data for the DLCA sample, as shown in Fig. 3. At small Q, unlike the RLCA sample, the data had an upward turn which could not be explained by the present model. The origin of this will be discussed shortly. However, if we use only the data above 0.002 Å^{-1}, a good fit is obtained with our model. The value of $D=1.73$ obtained for the fractal dimension agrees with previous measurements[6,7] and is consistent with the DLCA model in three dimensions.[4] Interestingly, we obtain $z_1=3.14$, which is substantially higher than the RLCA sample. The other fitted parameters were $z_2=4.86$ and $A_f=0.43$. As a test, we also fitted the RLCA data only using the points with $Q>0.002$ Å^{-1} to see if it significantly affected the fit, but it did not. A second DLCA sample was studied with similar results to the first except that its transmission factor for x rays was 200 times greater. This should rule out any multiple-scattering effects in these experiments.

An important result obtained by the fitting of our model to the data is that the first-shell coordination number for the DLCA sample is larger than for the RLCA sample, even though an aggregate prepared by DLCA is less dense.[6] It is known that in DLCA the potential between balls consists solely of an attractive van der Waals interaction, while in RLCA there is a Coulombic repulsion.[7] Presumably, then, in the early stages of the aggregation process, it is more favorable for a ball to stick to several balls in DLCA ($z_1>2$), but to only one ball in RLCA ($z_1\sim 2$). As the clusters get larger, however, these same properties in the potential will allow clusters formed by RLCA to interpenetrate more deeply, while in DLCA they will stick at the first contact, forming a less packed, more tenuous structure. These differences in first-shell coordination number can be expected to have large effects on the physical properties of the resultant aggregates, such as their conductivity and mechanical strength. Thus, our interpretation of these measurements provides insight into the early stages of the aggregation process and the effects of realistic interparticle potentials. Such interactions have not as yet been included

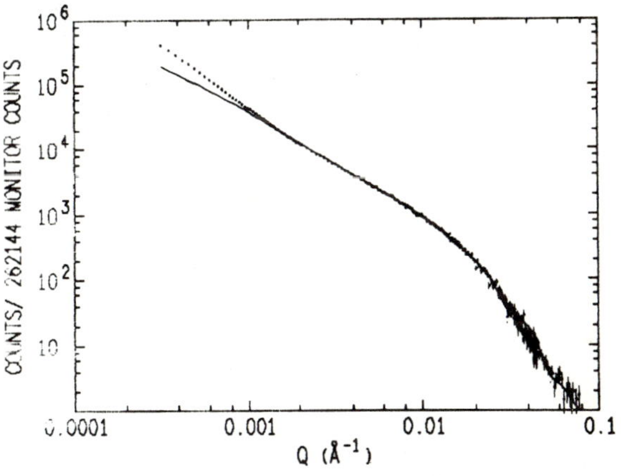

FIG. 3. Background-subtracted data of DLCA sample No. 1. The solid line is a fit to our model using only the data points above 0.002 Å^{-1}, but extended to smaller Q to show how the slope in the data increases there. The effective length scales and background level are the same as in Fig. 1.

VOLUME 57, NUMBER 5 PHYSICAL REVIEW LETTERS 4 AUGUST 1986

in any simulations.

Finally, we discuss the upward turn at small Q in the DLCA data. We believe that this reflects a restructuring of the aggregates,[11,15] which results in an apparent increase in the power law of the scattering at small Q, for distances greater than ~ 1000 Å. While this implies that the density is decreasing less rapidly than at shorter length scales, we cannot confirm that this large scale structure is fractal. It is significant that we observe this effect only for the DLCA samples which have lower fractal dimensions and are thus more tenuous and presumably structurally weaker than the RLCA sample. Further confirmation of this behavior was obtained from light-scattering measurements covering the range 2.5×10^{-4} Å$^{-1} \leqslant Q \leqslant 3\times10^{-3}$ Å^{-1}. Freshly prepared aggregates exhibited a clear fractal behavior with $D \sim 1.75$, while those that had been aged several days and subjected to shears similar to those of the x-ray samples showed an apparent fractal dimension of $D \sim 2.4$, which is roughly the same as the small-Q slope in the x-ray data. Similar observations have been reported for DLCA silica aggregates.[11] We note, however, that the x-ray data show that at short length scales the original fractal structure is preserved, while the restructuring occurs primarily at larger length scales.

In conclusion, we have shown that a more physical model of an aggregate which includes the short-range order is capable of accurately describing our data. The aggregates we have studied are consistent with a fractal interpretation over the length scales we have probed. However, it is crucial to include the short-range order to interpret the scattering data correctly. In so doing, it is also possible to extract information about the effects of interparticle interactions on the aggregation process.

It is a pleasure to acknowledge discussions with T. Witten, J. Gethner, R. Pynn, R. Ball, M. Y. Lin, M. O. Robbins, and W. D. Dozier. One of us (G.S.S.) acknowledges support from National Science Foundation Grant No. DMR-8307157. We would also like to thank the staff of Stanford Synchrotron Radiation Laboratory for their kind assistance.

[a]Also at Department of Physics, University of Colorado, Boulder, CO 80309.

[1]T. A. Witten and L. M. Sander, Phys. Rev. Lett. **47**, 1400 (1981), and Phys. Rev. B **27**, 5686 (1983).

[2]B. B. Mandelbrot, *The Fractal Geometry of Nature* (Freeman, San Francisco, 1982).

[3]P. Meakin, Phys. Rev. A **27**, 604, 1495 (1983).

[4]P. Meakin, Phys. Rev. Lett. **51**, 1119 (1983); M. Kolb, R. Botet, and R. Jullien, Phys. Rev. Lett. **51**, 1123 (1983); R. Jullien, M. Kolb, and R. Botet, J. Phys. (Paris), Lett. **45**, L211 (1984).

[5]M. Kolb and R. Jullien, J. Phys. (Paris), Lett. **45**, L977 (1984).

[6]D. A. Weitz and M. Oliveria, Phys. Rev. Lett. **52**, 1433 (1984).

[7]D. A. Weitz, J. S. Huang, M. Y. Lin, and J. Sung, Phys. Rev. Lett. **54**, 1416 (1985).

[8]D. A. Weitz, M. Y. Lin, J. S. Huang, T. A. Witten, S. K. Sinha, J. S. Gethner, and R. C. Ball, in *Scaling Phenomena in Disordered Systems*, edited by R. Pynn and A. Skjeltorp (Plenum, New York, 1985), p. 171.

[9]D. W. Schaefer, J. E. Martin, P. Wiltzius, and D. S. Cannell, Phys. Rev. Lett. **52**, 2371 (1984).

[10]S. K. Sinha, T. Freltoft, and J. Kjems, in *Kinetics of Aggregation and Gelation*, edited by F. Family and D. P. Landau (North-Holland, Amsterdam, 1984), p. 87; T. Freltoft, J. K. Kjems, and S. K. Sinha, Phys. Rev. B **33**, 269 (1986).

[11]C. Aubert and D. S. Cannell, Phys. Rev. Lett. **56**, 738 (1986).

[12]C. H. Bennett, J. Appl. Phys. **43**, 2727 (1972).

[13]P. Meakin, private communication.

[14]J. Turkevich, P. C. Stevenson, and J. Hillier, Discuss. Faraday Soc. **11**, 55 (1951).

[15]Y. Kantor and T. A. Witten, J. Phys. (Paris), Lett. **45**, L675 (1984).

VOLUME 54, NUMBER 10 PHYSICAL REVIEW LETTERS 11 MARCH 1985

Diffusion-Limited Aggregation in Two Dimensions

Alan J. Hurd and Dale W. Schaefer
Sandia National Laboratories, Albuquerque, New Mexico 87185
(Received 26 December 1984)

We have studied the aggregation of silica microspheres confined to two dimensions at an air-water interface. Under microscopic observation, both monomers and clusters are seen to aggregate by a diffusion-limited process. The clusters' fractal dimension is 1.20 ± 0.15, smaller than values obtained from current models of aggregation. We propose that anisotropic repulsive interactions account for the low dimensionality by more effectively repelling particles from the side of an existing dendrite than from the end.

PACS numbers: 64.60.Cn, 05.40.+j, 68.70.+w, 82.70.Dd

Colloid aggregation is the prototypical example of a random process which generates highly ramified dendritic structures. Although the qualitative features of aggregates have been well known, it was the recent discovery by Witten and Sander[1] of a computer simulation technique for diffusion-limited aggregation that spawned renewed interest in the structure of these curious objects. The Witten-Sander (WS) model and most subsequent modifications suggest that clusters possess universality in the sense that long-range structure is independent of the details of the interparticle potential. By now the number of theoretical results (including simulations) far exceeds the number of experimental results, and, since for practical reasons many simulations are carried out in two dimensions, measurements from real two-dimensional systems are particularly important.

Here we report the results of a two-dimensional aggregation experiment that we belive is the first reported physical realization of diffusion-limited aggregation in a plane, although the aggregation of nondiffusive floating wax particles[2] and the case of two-dimensional electrodeposition[3] have been reported. In our experiment, charged colloidal particles trapped by surface tension at the interface between air and water are allowed to aggregate by raising the salt concentration of the water substrate, thereby screening the repulsive electrostatic interactions. The aggregation process is studied by direct observation in an optical microscope.

In addition to the unusual two dimensionality, interfacial colloids provide a unique opportunity to characterize aggregation in detail. By direct observation one can gain insight into the controlling factors. Image-processing techniques can be applied to extract new information, such as the topological exponent[4] and the persistence length. Cluster-mass distributions and coagulation kinetics are also readily available. Internal dynamics as well as annealing processes can be observed directly or by light scattering. Finally, our experiment is not subject to the abiguities inherent in many others experiments such as the projection of a three-dimensional object onto a two-dimensional surface and the effects of gravity. In the present work we measure the fractal dimension and cluster-mass distribution; other properties are under study.

In situ scattering measurements of three-dimensional clusters, though scarce, currently offer the most reliable information for comparisons between competing models. Schaefer *et al.*[5] found that the fractal dimension D of colloidal silica aggregates, defined by $M \propto R_g^D$ where M is the mass and R_g the radius of gyration, is 2.1. The WS model, in which single diffusing particles aggregate onto a growing cluster, predicts $D = 2.5$[6] for three dimensions; the more realistic cluster-cluster (CLCL) model,[7,8] which allows many clusters to form and to aggregate among themselves, yields $D = 1.8$. Structures produced by the CLCL model are more "stringy" than those of the WS model because the colliding clusters cannot interpenetrate deeply. In both models, the particles are dilute and, until they stick together irreversibly on contact, noninteracting. The scattering experiments suggest that not all physical systems satisfy the assumptions of either model, perhaps most importantly with respect to the assumed absence of finite-range interactions.

In the presence of repulsive forces the Brownian motion of each particle cannot be considered independent; one would expect the motion to be biased away from existing clusters. The end of a charged dendrite therefore presents less of an electrostatic potential barrier than the side.[9] Although no simulations of diffusion-limited aggregation have yet appeared that include repulsive forces, anisotropic diffusion[10] and repulsive forces in an Eden model[11] have been simulated with the result that the fractal dimension decreases, consistent with structures having less branching. In our system of charged colloidal particles, finite-range forces would be expected to be important because the charge is largely responsible for the stability of the initial suspension.

The samples were prepared by spreading techniques outlined by Goodwin, Ottewill, and Parentich,[12] which were adapted for making small, controlled samples for microscopic examination [Fig. 1(a)]. A fresh suspension of silica microspheres (3000 Å diameter) with a

VOLUME 54, NUMBER 10 PHYSICAL REVIEW LETTERS 11 MARCH 1985

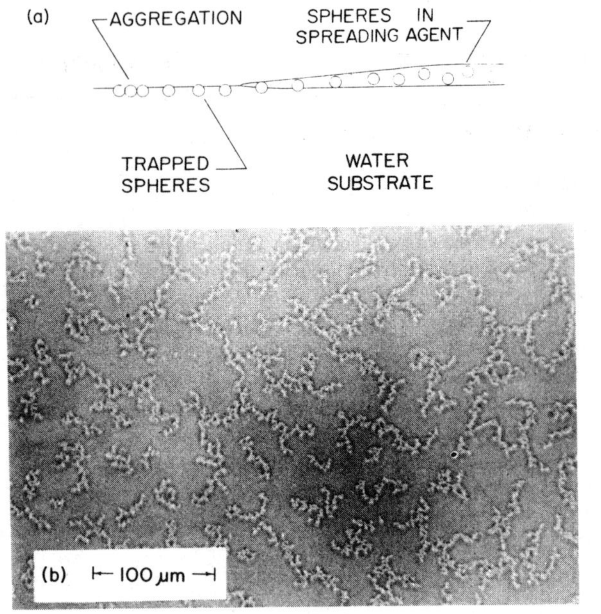

FIG. 1. (a) Schematic of aggregating interfacial colloids. A suspension of silica particles with a spreading agent is dispensed onto the surface of a salty-water substrate where the particles are trapped by surface tension. (b) Photomicrograph of actual two-dimensional clusters.

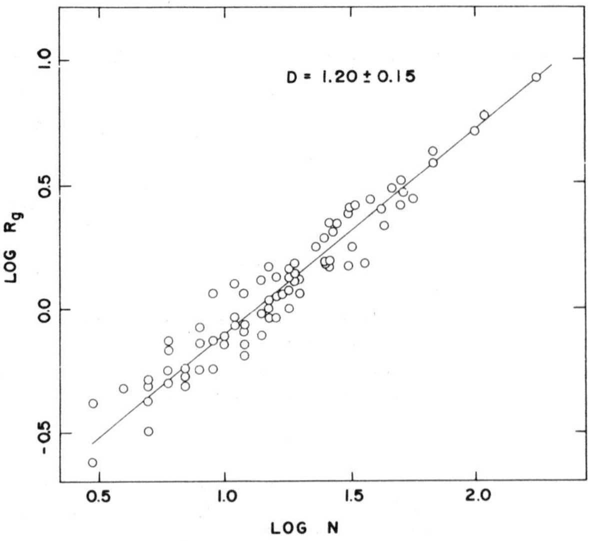

FIG. 2. Relationship between mass and radius of gyration. The units of R_j are arbitrary, and N is the number of equal-mass segments counted in the cluster, each containing about ten microspheres. The inverse slope (fractal dimension D) of the power law, which is followed for nearly two decades in mass, is 1.20—lower than the predictions given by computer simulations of noninteracting aggregating particles.

methanol spreading agent was dispensed onto a flat surface of "salty" water (1.0*N* CaCl) with a micrometer-driven syringe. Care was taken to avoid turbulence in the sample during spreading since the colloid is unstable to aggregation in the presence of high electrolyte concentrations.

Optical observations and photomicrographs were made as the sample proceeded to aggregate. Single particles could be seen sticking to each other and to developing clusters, but by the time reasonable cluster sizes were achieved (1 h) there were very few monomers left and the clustering of clusters was evident. The structures were digitized from photographs [Fig. 1(b)] by assigning a set of coordinates to each segment of equal mass as judged by eye, approximately ten microspheres per segment. The individual particles within a cluster could not be resolved in the photos because of their small size; larger, more resolvable spheres were found to take too long to aggregate. Thus a compromise was made in using small microspheres since the definition of the equal-mass segments during the digitization procedure was somewhat arbitrary. The scaling relation between the radius of gyration and the mass of a cluster is not greatly affected by this compromise, however, other than by the potential introduction of systematic errors and the loss of information about small length scales. To test for errors, the experiment was repeated with higher resolution and the results did not change significantly. In addition, several negatives were analyzed with image-processing equipment by digitizing the full area of a cluster image and then calculating the total area (mass) and second moment of the area (radius of gyration). Again, after accounting for the smearing of the optical image, the results did not change.

A plot of the logarithm of the radius of gyration R_g against the logarithm of the number of segments N (which is proportional to, but not equal to, the number of particles) is shown in Fig. 2 for 81 clusters ranging in size from 3 to about 150 segments (circa 30 to 1500 particles). Linear behavior holds over the decade and a half in mass that was measured. From the inverse slope, the fractal dimension is 1.20 ± 0.15, fitted with the data points weighted in favor of the larger clusters; this value is to be compared with 1.7 and 1.4, obtained for the two-dimensional WS and CLCL models, respectively.

Figure 3 illustrates the cluster-mass distribution $P(N)$. The most noteworthy feature is the existence of a well defined peak, also seen in CLCL simulations.[8] Evidently the mobility is an important factor in the aggregation rate since the smaller clusters are nearly depleted. We note that a scaling prediction for CLCL aggregation based on Smoluchowski's coagulation equation has the same qualitative form as $P(N)$.[13]

The low dimensionality of the clusters is consistent

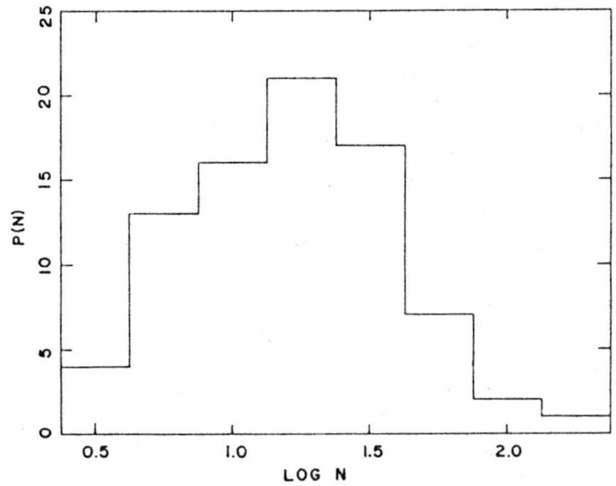

FIG. 3. Cluster-mass distribution. The well developed peak is a characteristic of both simulations and scaling predictions for the clustering of clusters (Ref. 13).

with their stringy appearance, which is quite different from WS clusters but similar to CLCL aggregates.[8] Guided by the work of Thomas and McCorkle on the formation of strings in the flocculation of thoria particles,[9] we analyzed the local details of the potential barriers surrounding the end of a dendrite. In the presence of repulsive electrostatic forces, the barrier encountered by a particle approaching a dendrite from the end is lower than the barrier for approaching the side: In the latter case, there are two or more particles in close proximity to a given growth site whereas at the end there is only one. Typically, the attractive van der Waals forces are not large enough to reverse this trend.

The particles are assumed to be trapped by surface-tension forces at the interface by the "flotation" mechanism commonly employed in many fields of chemical engineering, such as the removal of ore particles from suspension. Pieranski argues that the depth of the surface-energy potential well is several orders of magnitude greater than kT; therefore we can consider the particles to be confined strictly to two dimensions.[14] Moreover, since the potential well is very steep, gravitational forces can be neglected,[14] and the interface can be considered to be perfectly flat up to the particles' surfaces (no dimples). It is possible that significant deformations in the water surface around the spheres could be caused by electrical mechanisms, making surface-tension forces between spheres important. However, for simplicity we confine our attention to the electrostatic interactions.

On a dilute electrolyte substrate, the short-range interactions between two interfacial particles are dominated by a screened-Coulomb law according to Debye-Huckel theory.[15] For high concentration of electrolyte used in the present experiment, the use of a simple Debye-Huckel theory is not well justified. However, since the screening ions are repelled from the interface by their own image charges, the electrolyte is depleted near the interface in a zone roughly of thickness equal to the Debye screening length[16]; consequently the effective screening length is extended, and Debye-Huckel theory is more applicable. In any case, lack of a better theory forces us to use the simplest available expressions. Our estimates for a particle approaching a dimer show that the barrier for end-on approach is orders of magnitude lower than for side-on approach, an anisotropy whose magnitude can be attributed to the very high surface-charge density of colloidal silica. Thus, a growing aggregate has a strong tendency not to branch when there are large repulsive forces present.

Excluded-volume effects may also account for the low observed fractal dimension. Repulsive barriers expand the excluded volume (or, in the two-dimensional case, the excluded area) of each segment in a cluster. In flexible, equilibrium structures such as linear and branched polymers, excluded-volume effects lead to substantial swelling as the segments test various configurations by thermal motion.[17] We consider this effect to be secondary in importance because our structures are not very flexible, and they display distinct linearity on both long and short length scales, suggesting a bias in the growth process itself.

A similar possibility relates to the electrical contributions to the persistence length: a cluster with mutually repelling segments tends to straighten out. In the present experiment, however, the Debye screening length is smaller than the monomer diameter; therefore, we have rejected the electrical mechanism for stretching the clusters to a lower fractal dimension.

Finally, it should be realized that the clusters studied here may not be asymptotically large. The fact that the clusters seem to have a persistence length comparable to the size of the clusters themselves means that very large aggregates are required to see asymptotic exponents: The normal CLCL process requires at least 10^4 particles and, with biased growth[18] or a finite concentration[19] of particles in the bath, it probably requires more. Since many simulations are not carried out to the asymptotic limit, however, comparisons to them remain valid tests of our understanding. Moreover, our data show no sign of curvature, suggesting that, if there is a separate asymptotic regime, our results reflect the behavior for distances small compared to some unknown crossover length. Clearly the regime we have studied is an experimentally important one because it is readily accessible.

In summary, the differences between our results and extant models of aggregation indicate the importance of interparticle repulsive forces. Even though these are relatively short range, the local tendency to form strings is reflected in the global properties of the cluster.

This work was supported by the U. S. Department of Energy under Contract No. DE-AC04-76DP00789 and in part by Contract No. DE-FG02-84ER45084. Much of the work was carried out at the Martin Fisher School of Physics, Brandeis University, with the helpful suggestions and support of Seth Fraden and Bob Meyer. We are grateful to H. G. E. Hentschel and H. E. Stanley for helpful comments concerning the nonasympticity issue, and to Keith Keefer and Jim Martin for helpful comments.

[1]T. A. Witten, Jr., and L. M. Sander, Phys. Rev. Lett. **47**, 1400 (1981).

[2]C. Allain and B. Jouhier, J. Phys. (Paris) Lett. **44**, L421 (1983).

[3]M. Matsushita, M. Sano, Y. Hayakawa, H. Honjo, and Y. Sawada, Phys. Rev. Lett. 53, 286 (1984).

[4]D. C. Hong, S. Havlin, H. J. Herrmann, and H. E. Stanley, to be published; J. E. Martin, to be published.

[5]D. W. Schaefer, J. E. Martin, P. Wiltzius, and D. S. Cannell, Phys. Rev. Lett. **52**, 2371 (1984).

[6]P. Meakin, Phys. Rev. A **27**, 1495 (1983).

[7]P. Meakin, Phys. Rev. Lett. **51**, 1119 (1983).

[8]M. Kolb, R. Botet, and R. Jullien, Phys. Rev. Lett. **51**, 1123 (1983); R. Botet, R. Jullien, and M. Kolb, J. Phys. A Math. Gen. **17**, L75 (1984).

[9]I. L. Thomas and K. H. McCorkle, J. Colloid Interface Sci. **36**, 110 (1971).

[10]R. Jullien, M. Kolb, and R. Botet, J. Phys. **45**, 395 (1984).

[11]P. Meakin, Phys. Rev. B **28**, 6718 (1983).

[12]J. W. Goodwin, R. H. Ottewill, and A. Parentich, J. Phys. Chem. **84**, 1580 (1980).

[13]R. Jullien, M. Kolb, and R. Botet, in *Kinetics of Aggregation and Gelation,* edited by F. Family and D. P. Landau (North-Holland, Amsterdam, 1984), pp. 101–109.

[14]P. Pieranski, Phys. Rev. Lett. **45**, 569 (1980).

[15]F. H. Stillinger, J. Chem. Phys. **35**, 1584 (1961).

[16]L. Onsager and N. N. T. Samaras, J. Chem. Phys. **2**, 528 (1934).

[17]M. Daoud and F. Family, J. Phys. (Paris) **42**, 1359 (1981).

[18]H. E. Stanely, private communication.

[19]H. G. E. Hentschel and J. M. Deutsch, J. Chem. Phys. **81**, 2496 (1984).

VOLUME 52, NUMBER 12 PHYSICAL REVIEW LETTERS 19 MARCH 1984

Fractal Dimension of Dielectric Breakdown

L. Niemeyer, L. Pietronero,[a] and H. J. Wiesmann
Brown Boveri Research Center, CH-5405 Baden, Switzerland
(Received 23 November 1983)

It is shown that the simplest nontrivial stochastic model for dielectric breakdown naturally leads to fractal structures for the discharge pattern. Planar discharges are studied in detail and the results are compared with properly designed experiments.

PACS numbers: 77.50.+p, 02.50.+s

Dielectric breakdown of gaseous, liquid, and solid insulators frequently occurs by means of narrow discharge channels that exhibit a strong tendency to branching into complicated stochastic patterns.[1] Examples are lightnings,[2] surface discharges (Lichtenberg figures),[3] and treeing in polymers.[4] The global structure of branched discharges often shows a close structural similarity within a large class of discharge types but at the moment even a qualitative classification of these structures is missing.

In this Letter we present some evidence for fractal properties of branched discharges by analyzing adequately designed experiments and by the study of a new theoretical model. In particular we show that the simplest nontrivial stochastic model of dielectric breakdown naturally leads to fractal structures.

In order to facilitate the analysis of an experimental discharge pattern the optimal situation is that of a two-dimensional radial discharge like the one shown in Fig. 1. This example refers to a leader surface discharge (Lichtenberg figure) in compressed SF_6 gas whose properties have been studied in detail previously by Niemeyer and Pinnekamp.[5] The parameters were controlled in such a way that the experiment produces, to a good approximation, an equipotential channel system growing in a plane with a radial electrode from a central point.

If Fig. 1 corresponds to a fractal structure[6] the relation between the total length of all branches (or total number of points in the sense of Ref. 6) inside a circle of radius r and the radius r itself should be a power law with noninteger exponent D:

$$N(r) \sim r^D. \tag{1}$$

The thickness of the branches is considered as zero dimensional; it does not grow with the size of the object. Note that the apparent thickness in the photo of Fig. 1 is just an optic effect due to the number of carriers that have passed through a given branch. The number of branches $n(r)$ at a given distance r from the center is then given by

$$n(r) \sim dN(r)/dr \sim r^{(D-1)}. \tag{2}$$

Therefore a careful counting[7] of the number of branches for different values of r provides information about the exponent D. The analysis of the photos suggests a power-law behavior with $D \sim 1.7$, but the finite size and resolution do not allow an accurate determination of this exponent.

In order to gain insight into which features of the phenomenon of dielectric breakdown are the relevant ones with respect to its fractal properties we introduce an appropriate stochastic model and study it by computer simulations. We consider a two-dimensional square lattice (Fig. 2) in which a central point represents one of the electrodes while

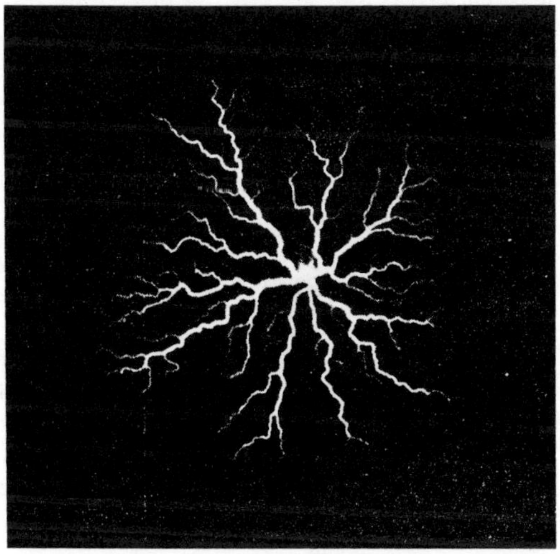

FIG. 1. Time-integrated photograph of a surface leader discharge (Lichtenberg figure) on a 2-mm glass plate in 0.3-MPa SF_6. Applied voltage pulse: 30 kV × 1 μs (Ref. 5). This experiment corresponds to an equipotential channel system growing in a plane with radial electrode.

VOLUME 52, NUMBER 12 PHYSICAL REVIEW LETTERS 19 MARCH 1984

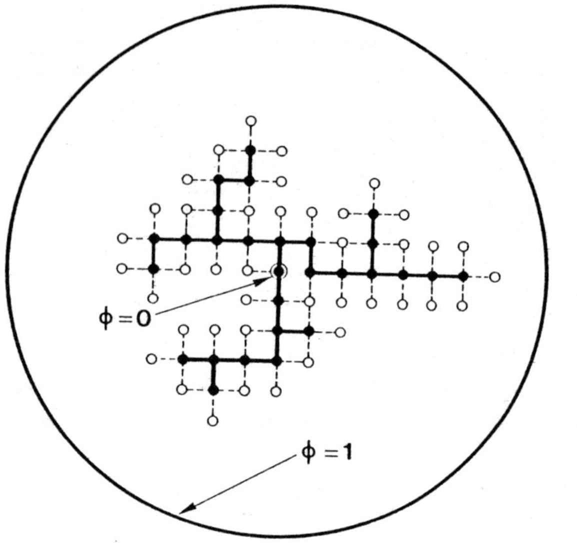

FIG. 2. Illustration of the stochastic model we introduce to simulate dielectric breakdown on a lattice. The central point represents one of the electrodes while the other electrode is modeled as a circle at large enough distance. The discharge pattern is indicated by the black dots connected with thick lines and it is considered equipotential ($\phi=0$). The dashed bonds indicate all the possible growth processes. The probability for each of these processes is proportional to the local electric field (see text).

the other electrode is modeled as a circle at large enough distance. The rules we assume for the growth of the discharge pattern are the following:

(a) The pattern grows stepwise. Figure 2 gives an example of a configuration after a few steps: The discharge pattern is indicated by the black dots connected with thick lines. The electric potential ϕ is defined for all points of the lattice by the discrete Laplace equation with the boundary conditions $\phi=0$ for each point of the discharge pattern and $\phi=1$ outside the external circle.

(b) At each step one bond is added to the pattern, linking a point of the pattern with a new point. The possible candidates are indicated in Fig. 2 by the dashed bonds that link a black point to a white one.

(c) To each of these dashed bonds a probability p is associated that is a function of the potential difference (local electric field[8]) between the black (i,k; $\phi_{ik}=0$) and white (i',k') dots connnected by this bond. The indices i,k and i',k' represent the discrete lattice coordinates. We write

$$p(i,k \rightarrow i',k') = \frac{(\phi_{i',k'})^\eta}{\sum(\phi_{i',k'})^\eta}, \quad (3)$$

where a power-law dependence with exponent η is assumed to describe the relation between local field and probability. The sum in the denominator refers to all the possible growth processes (dashed lines in Fig. 2). Given this probability distribution a new bond (and point) is chosen randomly and added to the discharge pattern. With this new discharge pattern one starts again. These rules properly define also the starting of the process from the central point. Further it follows that no crossing is possible and that the pattern is simply connected.

The essence of this stochastic model is therefore that the growth probability depends on the local field (potential) determined by the equipotential discharge pattern. The most laborious part of this program is the solution, at each time step, of the Laplace equation

$$\nabla^2\phi = 0. \quad (4)$$

The discrete form of Eq. (4) on the two-dimensional lattice[9] can be written as

$$\phi_{i,k} = \tfrac{1}{4}(\phi_{i+1,k} + \phi_{i-1,k} + \phi_{i,k+1} + \phi_{i,k-1}). \quad (5)$$

Given the appropriate boundary conditions the potential is obtained by iterating Eq. (5). Typically good convergence is obtained with a number of iterations between 5 and 50. This method correctly reproduces the global influence of a given discharge pattern on the growth probability for each bond. So for example, the tip of a line like the one on the right side of Fig. 2 will have a large growth probability while the points inside a cage (e.g., left side of Fig. 2) will have much smaller probability. These are the well known "tip effect" and "Faraday screening" that result from the solution of the field equation.

Recently Sawada *et al.*[10] have tried to simulate dielectric breakdown by simply assigning *a priori* a larger probability for the growth of tips with respect to side branching. This simplification completely neglects the nonlocal nature of the electric field. For example a tip inside a Faraday cage has in their approach the same growth probability as a free tip. The results obtained with this approximation are qualitatively different from experimental discharges and their global dimensionality remains the Euclidean one ($D=2$). Badaloni and Gallimberti[11] instead do not approach the problem of the dimensionality of the structure but provide an interesting connection between a microscopic picture of the avalanche formation and the bifurcation probability and geometry at a tip.

We start the discussion of the computer simulations with the case $\eta=1$ (growth probability proportional to the local field) that is the most realistic case for the present experiment.[5] We obtain highly branched structures like the one shown in Fig. 3. In order to investigate the dimensionality of these discharge structures we have generated several of them and we have plotted the logarithm of the total number of points within a certain radius as a function of the logarithm of this radius. The average over five large samples of about 5000 points each gives rise to a Hausdorff dimension $D = 1.75 \pm 0.02$. The error bar is due to the statistical fluctuations of the slope but the possibility of a larger systematic error due to the finite size of the systems considered cannot be excluded. We can see that the computed value is in good agreement with the value $D \simeq 1.7$ suggested by the analysis of the experimental discharges. This comparison anyhow should be considered with caution because in the computer simulations we have two-dimensional field equations while in the experiment only the growth is two-dimensional while the field lines extend three dimensionally. This is a point that we intend to consider in more detail in future work.

We study further the effect of values of η different from 1. This is of interest because in other systems than gases (solids, liquids, and polymers) the microscopic relation between growth probability and local field may be more appropriately described by a nonlinear function.[1,4]

The case $\eta = 0$ corresponds to growth probabilities independent on the local field. This is a sort of Eden (cancer) model[12] with the difference that

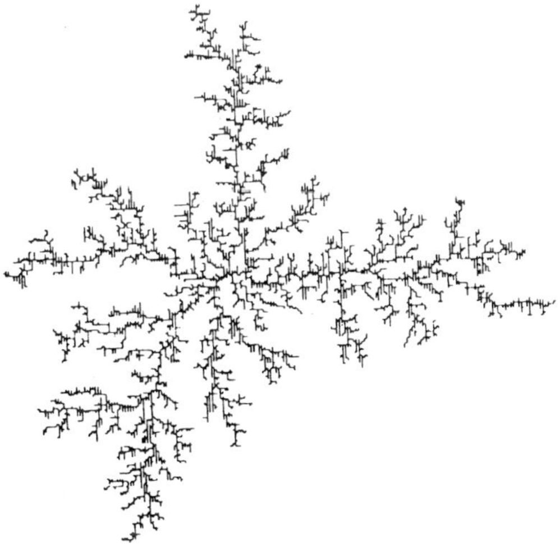

FIG. 3. Example of computer-generated discharge pattern of about 5000 steps. The Hausdorff dimension for these structures turns out to be $D = 1.75 \pm 0.02$.

some of the white points in our case (Fig. 2) have a higher probability because they can be reached from more than one black point. The growth is in this case homogeneous with $D = 2$. We have further considered the cases $\eta = 0.5$ and $\eta = 2$. The corresponding Hausdorff dimensions are reported in Table I and they differ from the case $\eta = 1$. This shows that the Hausdorff dimension depends on the parameter η and there is no universality in this respect. For large η the structure tends to be more "linear," but a careful study of the limit $\eta \to \infty$ is not possible with the present method. We suggest, however, that, in contrast with the simple behavior for $\eta \to 0$, the limit $\eta \to \infty$ is likely to be nonanalytical in the sense that the exchange of limits $\eta \to \infty$ and $N \to \infty$ may give rise to different behaviors.[14]

Another interesting variation we have considered is the effect of a Debye screening (due to eventual mobile carriers in the system) or correspondingly a finite mass in a field theory. This is realized by replacing ∇^2 in Eq. (4) by $(\nabla^2 - k^2)$. Such a change introduces a length scale $(l \sim k^{-1})$ into the system that then loses its self-similarity. The results show nonhomogeneous patterns for length scales lower tha l but homogeneous behavior $(D = 2)$ on larger scales.

With respect to other growth models the general shape of our patterns for $\eta = 1$ (Fig. 3) and the corresponding value of D closely recall the results of the diffusion-limited aggregation models.[12,15-17] This similarity actually arises from the close relations between random walk and potential theory, a point to which we intend to dedicate a more extended discussion elsewhere.

In summary we have introduced a new stochastic model to describe the discharge pattern of dielectric breakdown. The basic assumption is that the growth probability depends on the local electric field. We show that this model naturally leads to fractal structures that we study in detail in two

TABLE I. Dependence of the Hausdorff dimension D on the exponent η used in the relation between probability and local field [Eq. (3)].

η	D
0	2
0.5	1.89 ± 0.01
1	1.75 ± 0.02
2	~ 1.6[a]

[a]Ref. 13.

VOLUME 52, NUMBER 12 PHYSICAL REVIEW LETTERS 19 MARCH 1984

dimensions. The results are consistent with the analysis of properly designed experiments of planar discharge.

It is a pleasure to thank J. Bernasconi, T. Hibma, L. Peliti, P. Pfluger, W. R. Schneider, and H. R. Zeller for interesting discussions, F. Pinnekamp for supplying the discharge photographs, and M. Dietiker for help in the computer graphics. One of the authors (L.P.) is also grateful for the hospitality at the Institute for Theoretical Physics of the University of Santa Barbara where, sharing an office with P. Meakin, he learned about the models of diffusion-limited aggregation.

---

[a]Address from April 1984: Solid State Physics Laboratory, University of Groningen, Melkweg 1, 9718EP-Groningen, The Netherlands.

[1]*Engineering Dielectrics, Vols. I–VI* (American Society for Testing and Materials, Philadelphia, 1983), Special Technical Publication No. LC No82-70637.

[2]J. M. Meek and J. D. Craggs, *Electrical Breakdown of Gases* (Wiley, New York, 1978). Atmospheric lightnings are actually rather complicated to analyze because of the nonhomogeneity of the density, humidity, and conductivity of the air.

[3]E. Nasser, *Fundamentals of Gaseous Ionization and Plasma Electronics* (Wiley, New York, 1971).

[4]R. A. Fava, in *Treatise on Material Science and Technology*, edited by J. M. Schutz (Academic, New York, 1977), Vol. 10, Pt. B, p. 677.

[5]L. Niemeyer and F. Pinnekamp, in *Gaseous Dielectrics III*, edited by L. G. Christophorou (Pergamon, New York, 1982), p. 379.

[6]B. Mandelbrot, *Fractals: Form, Chance and Dimension* (Freeman, San Francisco, 1977).

[7]We have counted all the branches visible in the original negative. Some tiny ones could disappear in the printed reproduction of the photo.

[8]The probabilistic approach reproduces the stochastic nature of the ionization avalanche formation that starts the discharge progression.

[9]Actually in the experiment only the discharge is planar while the field extends three dimensionally.

[10]Y. Sawada, S. Ohta, M. Yamazaki, and H. Honjo, Phys. Rev. A 26, 3557 (1982).

[11]S. Badaloni and I. Gallimberti, in *Proceedings of the Eleventh International Conference on Phenomena in Ionized Gases, Prague, Czechoslovakia, 1973*, edited by I. Stoll (Czechoslavakia Academy of Sciences, Prague, Czechoslovakia, 1973), p. 196.

[12]For an overview on various models of aggregation see, H. E. Stanley, F. Family, and H. Gould, to be published.

[13]The case $\eta = 2$ gives rise to patterns with scarce ramifications for which one should probably consider larger sizes in order to obtain a reliable value for D.

[14]An example of this phenomenon can be found in the "true" self-avoiding walk: L. Pietronero, Phys. Rev. B 27, 5887 (1983).

[15]T. A. Witten and L. M. Sander, Phys. Rev. Lett. 47, 1400 (1981).

[16]T. A. Witten and L. M. Sander, Phys. Rev. B 27, 5686 (1983). These authors have already pointed out a possible similarity between diffusion-limited aggregation and dielectric breakdown.

[17]P. Meakin, Phys. Rev. A 27, 1495 (1983).

VOLUME 55, NUMBER 9 PHYSICAL REVIEW LETTERS 26 AUGUST 1985

Occupancy-Probability Scaling in Diffusion-Limited Aggregation

Leonid A. Turkevich and Harvey Scher

Corporate Research Center, The Standard Oil Company, Cleveland, Ohio 44128

(Received 8 April 1985)

A continuous-time random-walk theory of diffusion-limited aggregation yields perimeter occupancy probabilities. Scaling relates the fractal dimension D to the cluster-tip occupancy probabilities. These agree with the analytic probabilities near cusps of a lattice-symmetric array of traps. On a two-dimensional square lattice $D=\frac{5}{3}$, whereas $D=2$ for the Eden model, and $D=\frac{4}{3}$ for the $\eta=2$ dielectric breakdown model. D is not universal: $D=\frac{7}{4}$ for the two-dimensional triangular lattice. The square and triangular lattices bracket ($\pm 2.5\%$) Meakin's large off-lattice simulations ($D=1.71$).

PACS numbers: 82.70.Dd, 05.40.+j, 64.60.Cn, 68.70.+w

The irreversible aggregation of colloids[1] and aerosols[2] is often rate limited by diffusion of the particles to the surface of the aggregate. Similar dendritic patterns[3] arise in chemical precipitation from a supersaturated solution or upon crystallization from a supersaturated melt.[4] The multibranched aggregates so formed have been attributed[4] to short-wavelength growth instabilities inherent in these diffusion-limited processes.

Witten and Sander[5,6] have devised a lattice model which simulates such diffusion-limited aggregation (DLA): A cluster is grown by successive accretion of random walkers to perimeter sites. This DLA model is sufficiently simple as to be amenable to numerical simulation.[5–7] These computer experiments have established scale invariance for the clusters and have characterized them by a Hausdorff (fractal) dimension D.

Witten and Sander[6] argue against an upper critical dimensionality for DLA. Several conjectures[7,8] attempt to relate D to the Euclidean dimension d in which the cluster is embedded. While inconclusive as to establishing D, renormalization-group arguments[9] suggest that DLA is in a universality class distinct from both equilibrium lattice animals and the unrestricted random walk. This Letter addresses the theoretical questions of universality and the Hausdorff dimension of DLA clusters.

We first reformulate DLA as a continuous-time random walk (CTRW).[10] Our formal solution for a given cluster yields the perimeter occupancy probabilities. We utilize this probability distribution to generate DLA clusters, without simulating the entire lattice random walk.

Scaling identifies D from the occupancy probability for the maximally extending portion of the cluster. The DLA cluster is thus characterized not by its complicated random interior but rather by the simple growth of its perimeter. We argue that the occupancy probabilities for the unscreened DLA cluster tips may be obtained from occupancy probabilities for similar tips of regular (lattice-symmetric) clusters. The fundamental feature of the growing boundary is its cusp structure; this feature is entirely reproduced by the simpler geometry. Application of the analytic occupancy probabilities for a square to DLA on a two-dimensional square lattice yields $D=\frac{5}{3}$; we similarly obtain $D=\frac{7}{4}$ for a 2D triangular lattice.

For specificity, consider a 2D square lattice. Let $R(\mathbf{s},t)$ be the probability (per unit time) for a random walker just to have arrived at a site $\mathbf{s}$ at time t. Let $\psi(\mathbf{s},\mathbf{s}';t)$ be the joint probability (per unit time) both that the time interval between successive arrivals is t and that the displacement that occurs is from $\mathbf{s}'$ to $\mathbf{s}$. We consider an intermediate stage where the lattice contains N contiguously occupied sites. We distinguish perimeter sites $\mathbf{p}\in\Pi$, which are perfectly absorbing traps, from all other lattice sites, which possess nearest-neighbor transition rates W:

$$\psi(\mathbf{s},\mathbf{s}',t)=\begin{cases}0, & \mathbf{s}'\in\Pi,\\ W(\delta_{\mathbf{s},\mathbf{s}'\pm a\hat{\mathbf{x}}}+\delta_{\mathbf{s},\mathbf{s}'\pm a\hat{\mathbf{y}}})e^{-4WT}, & \mathbf{s}'\notin\Pi,\end{cases} \tag{1}$$

where a is the lattice constant. The Laplace transform of $R(\mathbf{s},t)$ satisfies

$$\tilde{R}(\mathbf{s},u)=\sum_{\mathbf{s}'\notin\Pi}\tilde{\psi}(\mathbf{s}-\mathbf{s}';u)\tilde{R}(\mathbf{s}',u)+\delta_{\mathbf{s},\mathbf{s}_0}, \tag{2}$$

with an initial site $\mathbf{s}_0$ for the random walker. Formally,[11]

$$\tilde{R}(\mathbf{s},u)=\tilde{G}(\mathbf{s}-\mathbf{s}_0,u)-\sum_{\mathbf{p}\in\Pi}[\tilde{G}(\mathbf{s}-\mathbf{p},u)-\delta_{\mathbf{s},\mathbf{p}}]\tilde{R}(\mathbf{p},u), \tag{3}$$

VOLUME 55, NUMBER 9 PHYSICAL REVIEW LETTERS 26 AUGUST 1985

where the Green's function for the perfect lattice is given by

$$\tilde{G}(\mathbf{s}-\mathbf{s}_0,u)=\frac{1}{\pi^2}\int_0^{\pi}d\theta_x\int_0^{\pi}d\theta_y\frac{\cos m\,\theta_x\cos n\,\theta_y}{1+(1/2\sigma)(\cos\theta_x+\cos\theta_y)},\tag{4}$$

with $\sigma=1+u/4W$ and $\mathbf{s}-\mathbf{s}_0=ma\,\hat{\mathbf{x}}+na\,\hat{\mathbf{y}}$. Morita's recursion relations[12] facilitate use of these lattice Green's functions. Evaluating (3) for a perimeter site $\mathbf{p}$, we have

$$\tilde{G}(\mathbf{p}-\mathbf{s}_0,u)=\sum_{\mathbf{p}'\in\Pi}\tilde{G}(\mathbf{p}-\mathbf{p}',u)\tilde{R}(\mathbf{p}',u).\tag{5}$$

The probability that the $(N+1)$st random walker ultimately lands at $\mathbf{p}$ is given by $P(\mathbf{p})=P(\mathbf{p},t\to\infty)=\tilde{R}(\mathbf{p},u\to 0)$, and so we need only invert (5) in the limit $u\to 0$. Thus the CTRW *exactly solves* the DLA growth problem: Given a cluster, we obtain the probability for the next random walker to occupy each perimeter site.

We have incorporated this exact solution into an algorithm for generating DLA clusters, without simulating the entire random walk. Our specific realization of the random walk is accomplished by inverting $R=C(\mathbf{p})$, where R is a random number and $C(\mathbf{p})$ is the cumulant of $P(\mathbf{p})$. In this way, we iteratively build up the cluster; an example $(N=150)$ is shown in Fig. 1. The method becomes unwieldly for larger clusters, because of the matrix inversion of (5), the matrix dimensions being proportional to the number of perimeter sites.[13] Also shown in Fig. 1 are the probabilities for subsequent occupancy of the perimeter sites. We emphasize that while our clusters are not large, this method provides the additional information of the occupancy probabilities, which, in the usual technique of brute-force simulation, can only be achieved by many realizations *at each stage of the growth.*

We evaluate D for our clusters from the number of occupied sites within a radius r about the center of mass: $N_{\text{occ}}(r)\sim(r/a)^D$. For the square lattice we find $D\simeq 1.7$, consistent with the largest simulations.[7] Algebraic behavior obtains for $N\gtrsim 20$.

Our solution for the occupancy probabilities motivates a scaling argument for D. As the cluster interior is screened[6] by its exterior branches [screening length ξ^N], interior perimeter sites have negligible probability of subsequent occupancy. Consistent with this screening, the DLA cluster cannot be "fully developed" to its tips: Near the tips, the cluster has fewer occupied sites than it would were it fully developed to its nominal radius $r_0^{(N)}=N^{1/D}a$. As it is fully developed only within a radius $r_-^{(N)}=r_0^{(N)}-\xi^{(N)}$, it must extend further, to a distance[14] $r_+^{(N)}>r_0^{(N)}$ (Fig. 2). In what follows, we use only that $r_0^{(N)}$, $\xi^{(N)}$, and $r_+^{(N)}$ scale similarly with N, which scaling is guaranteed by self-similarity.

We now show that the occupancy-probability density $p_N(r)$ determines D. Given an N cluster, the probability that the $(N+1)$st random walker will land at the outermost portion of the cluster [the annulus at $r_+^{(N)}$] is

$$P_{\max}=\int_{r_+^{(N)}-a}^{r_+^{(N)}}p_N(r)dr,\tag{6}$$

in which case the cluster grows: $r_+^{(N+1)}=r_+^{(N)}+a$. If the $(N+1)$st random walker lands deeper within the cluster [but still in the edge region $r_-^{(N)}<r<r_+^{(N)}$], the cluster does not grow: $r_+^{(N+1)}=r_+^{(N)}$. Combining

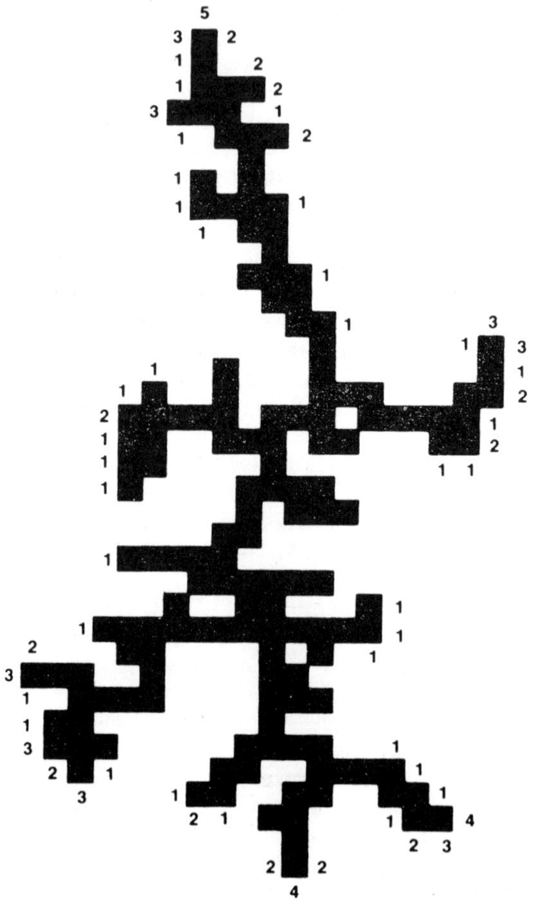

FIG. 1. A cluster $(N=150)$ grown according to the CTRW algorithm. Shown also are occupancy probabilities (in percent) for all perimeter sites with $P(\mathbf{p})>0.005$. Appreciable occupancy probability occurs only at the tips of the cluster; the interior is screened.

VOLUME 55, NUMBER 9 PHYSICAL REVIEW LETTERS 26 AUGUST 1985

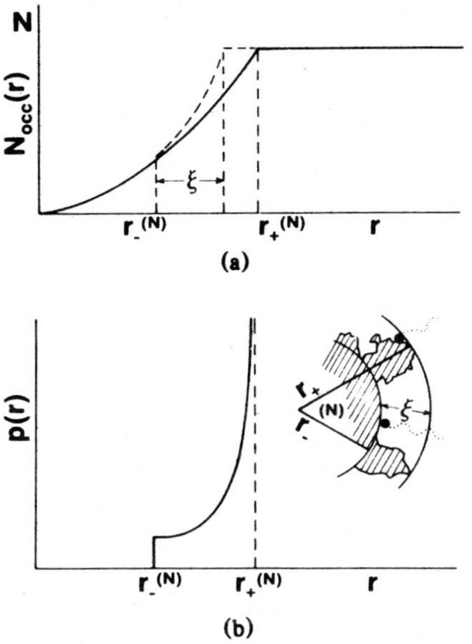

FIG. 2. (a) Schematic of the number of occupied sites $N_{occ}(r)$ within distance r from the center of mass. The cluster is "fully developed" $[N_{occ}(r) \sim (r/a)^D]$ for $r < r_-^{(N)}$ with a lower-density "edge" region for $r_-^{(N)} < r < r_+^{(N)}$. A cluster with no edge region would extend to a nominal radius $r_0^{(N)}$. All growth takes place in this edge region. (b) Schematic of the occupancy probability density $p_N(r)$ in the edge region. No growth occurs for $r < r_-^{(N)}$: $p_N(r)=0$. $p_N(r)$ is singular at $r_+^{(N)}$, reflecting maximal growth probability at the tips. Inset: Two possible random walkers, one adhering near a cluster tip $(r_+^{(N)})$, and the other adhering deep within the edge region $(r_-^{(N)})$.

these alternatives,

$$r_+^{(N+1)} = P_{max}(r_+^{(N)} + a) + (1 - P_{max}) r_+^{(N)}.$$

For large N,

$$dr_+^{(N)}/dN = P_{max}(N)a. \tag{7}$$

The growth of the cluster is thus driven by the occupancy probability of the tips. If particles are added at the tips, the cluster grows; if particles are added to the interior, the cluster does not grow. With cluster growth, the occupancy probability for the next random walker is altered; otherwise the next random walker experiences essentially the same perimeter of traps. Hence (7) also self-consistently expresses the change in occupancy-probability density as particles are added to the cluster. Since $r_+^{(N)} \sim N^{1/D}a$, (7) and the divergence of $p_N(r)$ at r_+^N determine D.

As we require $p_N(r)$ only at the tips, the intricacies of the random, ramified cluster interior are irrelevant—*all the growth takes place at the tips.* We thus may utilize *any* object possessing the same cusp structure as a DLA cluster. This is the fundamental

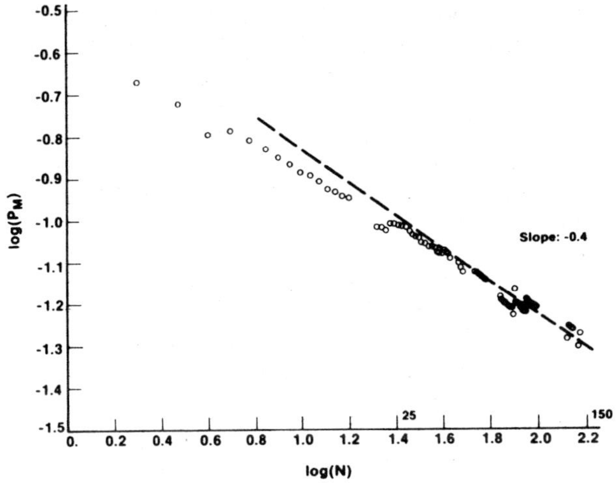

FIG. 3. $P_{max}(N)$ for the growing cluster of Fig. 1, scaling as predicted.

feature of our approach to the probability scaling: D does not depend on the cluster's random, irregular structure but rather on *how it grows.* We can thus utilize a regular object with the symmetry of the lattice—for the square lattice, a square with sides $2L$—i.e., we consider a random walker in the presence of a square array of absorbing traps. We compute the occupancy probabilities as a function of distance δs away from the corners of the square.

For $t \to \infty$, the random-walk diffusion problem reduces to Laplace's equation with perfectly absorbing walls. This is equivalent[15] to the electrostatic problem $\mathbf{E} = -\nabla\phi$ of a conducting square held at $\phi = 0$ with a conducting circle at infinity held at $\phi = 1$. The probability density,

$$p(\mathbf{r}) = |\mathbf{E}(\mathbf{r})| / \int_\Pi |\mathbf{E}(\mathbf{r})| ds, \tag{8}$$

is normalized over the perimeter Π of the square.

The electrostatics may be solved exactly via conformal transformation, $E_\perp \sim 1/\mathrm{cn}\zeta$, where the variable ζ is given implicitly by

$$z = x + iy = \frac{-iL}{E - K/2}\left[\mathrm{zn}\zeta + \left(\frac{E}{K} - \frac{1}{2}\right)\zeta\right] + L. \tag{9}$$

Here $\mathrm{cn}\zeta$ and $\mathrm{zn}\zeta$ are Jacobi elliptic functions, and K and E are the usual complete elliptic integrals, all of modulus $k = 1/\sqrt{2}$. Near the tips ($\delta s \to 0$), the probability density (8) diverges as

$$p(\delta s) \sim \frac{(E - K/2)^{2/3}}{\sqrt{2}\pi L}\left(\frac{L}{12\delta s}\right)^{1/3}. \tag{10}$$

We assert that the scaling of the probability density near a square-lattice DLA tip is reproduced by that of the square near a corner,

$$p_N(r) \sim \tfrac{2}{3}\zeta^{(N)-2/3}(r_+^{(N)} - r)^{-1/3}, \tag{11}$$

normalized over the screening length $\zeta^{(N)}$. We verify this assertion below (Fig. 3). With use of (11) in (6), $P_{\max} \sim (a/\zeta)^{2/3}$. Substitution into (7) gives $N \sim (r/a)^{5/3}$, i.e., $D = \frac{5}{3}$ for 2D DLA on a square lattice.

We verify the consistency of using (11) to obtain the singular behavior of the DLA probability density near the cluster tips. Figure 3 displays $P_{\max}(N)$ as the cluster of Fig. 1 is grown. Even for such small clusters, $P_{\max} \sim \zeta^{-2/3} \sim N^{-2/5}$. Similar scaling ($P_{\max} \sim N^{-0.39}$) is observed by Meakin[16] in averages over many (10^5) simulations of large clusters ($N \sim 5 \times 10^4$).

We remark that (8) may be trivially extended,

$$p(\mathbf{r}) = |\mathbf{E}(\mathbf{r})|^{\eta} / \int_{\Pi} |\mathbf{E}(\mathbf{r})|^{\eta} \, ds, \qquad (12)$$

to yield $D = 2$ for the Eden cancer model[17] ($\eta = 0$) and $D = \frac{4}{3}$ for the $\eta = 2$ Brown-Boveri dielectric breakdown model[15] on a 2D square lattice.

We have investigated the lattice dependence of the above results. The random-walk problem on a 2D triangular lattice can also be solved, where we utilize a hexagonal array of traps to obtain the probability density near a DLA cluster tip,

$$p_N(r) \sim \tfrac{3}{4} \xi^{(N)-3/4} (r_+^{(N)} - r)^{-1/4}. \qquad (13)$$

Use of (13) in (6) and substitution into (7) gives $D = \frac{7}{4}$ and $P_{\max} \sim N^{-3/7}$. The results for fourfold- ($D = \frac{5}{3}$) and sixfold-coordinated ($D = \frac{7}{4}$) lattices bracket and are within $\pm 2.5\%$ of the latest[18] numerical results ($D = 1.71$) for DLA *without* a lattice, where we expect averaging over coordination.

Finally the random-walk problem on *anisotropic* fourfold-coordinated lattices (2D oblique lattices with arbitrary angle of inclination β) can also be solved by conformal transformation. Near the sharp corner,

$$p_N(r) \sim [\pi/(2\pi - \beta)] \xi^{(N)-x} (r_+^{(N)} - r)^{x-1}, \qquad (14)$$

with $x = \pi/(2\pi - \beta)$. In this case, $D = (3\pi - \beta)/(2\pi - \beta)$; while nonuniversal, D is stringently bracketed: $1.50 < D < 1.67$. Numerical simulations on these anisotropic lattices will provide a strong test of the theory.

In summary, we have presented a CTRW reformulation of DLA. Our solution yields, at each stage of the growth, the perimeter occupancy probabilities of the random walker. Growth occurs predominantly at the cluster tips, and is controlled by the occupancy probability $P_{\max}$ of these maximally extending tips. The scaling of $P_{\max}$ determines the Hausdorff dimension D of the cluster, and may be obtained by use of the singular part of the probability density for a regular object with the same cusps as exhibited by the DLA cluster tips. We find $D = \frac{5}{3}$ for DLA on a 2D square lattice and $D = \frac{7}{4}$ on a 2D triangular lattice, thus evincing nonuniversality. We have verified the predicted scaling, $P_{\max} \sim N^{-2/5}$, for the square lattice.

We thank J. C. Gray for assistance in the numerical generation of the clusters.

---

[1]B. V. Enüstün and J. Turkevich, J. Am. Chem. Soc. **85**, 3317 (1963); M. J. Vold, J. Colloid Sci. **18**, 684 (1963); A. I. Medalia, Surf. Colloid Sci. **4**, 1 (1971); D. A. Weitz and M. Oliveria, Phys. Rev. Lett. **52**, 1433 (1984).

[2]S. K. Friedlander, *Smoke, Dust and Haze* (Wiley, New York, 1977).

[3]J. S. Langer and H. Müller-Krumbhaar, Acta Metall. **26**, 1681, 1689 (1978); H. Müller-Krumbhaar and J. S. Langer, Acta Metall. **26**, 1697 (1978).

[4]W. W. Mullins and R. F. Sekerka, J. Appl. Phys. **34**, 323 (1963).

[5]T. A. Witten, Jr., and L. M. Sander, Phys. Rev. Lett. **47**, 1400 (1981).

[6]T. A. Witten, Jr., and L. M. Sander, Phys. Rev. B **27**, 5686 (1983).

[7]P. Meakin, Phys. Rev. A **27**, 604, 1495 (1983).

[8]M. Muthukumar, Phys. Rev. Lett. **50**, 839 (1983); M. Tokuyama and K. Kawasaki, Phys. Lett. **100A**, 337 (1984); H. G. E. Hentschel, Phys. Rev. Lett. **52**, 212 (1984).

[9]H. Gould, F. Family, and H. E. Stanley, Phys. Rev. Lett. **50**, 686 (1983).

[10]E. W. Montroll and G. Weiss, J. Math. Phys. **6**, 167 (1965); H. Scher and M. Lax, Phys. Rev. B **7**, 4491 (1973).

[11]E. W. Montroll and H. Scher, J. Math. Phys. **9**, 101 (1973).

[12]T. Morita, J. Math. Phys. **12**, 1744 (1971).

[13]We have reduced the matrix dimension by including a pattern-recognition algorithm to eliminate perfectly screened perimeter sites.

[14]With the assumption of a power-law density in the edge region, with continuity at $r_-^{(N)}$, $\rho_{\mathrm{edge}}(r) \sim \rho_{\mathrm{fd}}(r_-^{(N)}) \times (r_-^{(N)}/r)^{\alpha}$, with the fully developed density $\rho_{\mathrm{fd}}(r) \sim (D/2\pi a^2)(r/a)^{D-2}$, and $\alpha > 2 - D$. This determines $r_+^{(N)} \sim r_0^{(N)} + (D - 2 + \alpha)\xi^{(N)2}/2r_0^{(N)}$.

[15]L. Niemeyer, L. Pietronero, and H. J. Wiesmann, Phys. Rev. Lett. **52**, 1033 (1984).

[16]P. Meakin, private communication.

[17]M. Eden, in *Proceedings of the Fourth Berkeley Symposium on Mathematical Statistics and Probability,* edited by J. Neyman (Univ. of California Press, Berkeley, 1961), Vol. 4, p. 223; H. P. Peters, D. Stauffer, H. P. Hölters, and K. Loewenich, Z. Phys. B **34**, 399 (1979).

[18]P. Meakin, Bull. Am. Phys. Soc. **30**, 222 (1985).

Reprinted with permission from *Nature* **314**, J. Nittmann, G. Daccord, and H.E. Stanley, "Fractal Growth of Viscous Fingers: Quantitative Characterization of Fluid Instability Phenomenon," pp. 141–144, ©1985 Macmillan Magzines Ltd., London.

Fractal growth of viscous fingers: quantitative characterization of a fluid instability phenomenon

Johann Nittmann*, Gérard Daccord* & H. Eugene Stanley†

* Etudes et Fabrication Dowell Schlumberger 42003 St Etienne, France
† Center for Polymer Studies and Department of Physics, Boston University, Boston, Massachusetts 02215, USA

What happens when one attempts to push water through a fluid of higher viscosity? Under appropriate experimental conditions, the water breaks through in the form of highly branched patterns called viscous fingers. Water was used to push a more viscous but miscible, non-newtonian fluid in a Hele-Shaw cell. The resulting viscous finger instability was found to be a fractal growth phenomenon. Reproducible values of the fractal dimension d_f *were found and were interpreted using a modification of the diffusion limited aggregation model.*

WHEN one attempts to use a low viscosity fluid (like water) to push high viscosity fluid (like oil), there arises a fluid instability phenomenon termed the 'viscous fingering instability' in which the low viscosity fluid forms characteristic fingers extending well into the high viscosity fluid[1–4]. In addition to its obvious interest in enhanced oil recovery by water flooding, and other practical applications, this phenomenon is of considerable basic interest and is not understood at present. It may be studied in detail in the laboratory using a Hele-Shaw cell, which is a pair of transparent plates separated by typically 0.5 mm. The space between the plates is filled with high viscosity fluid and the low viscosity fluid is injected. Normally, the interfacial tension between the two fluids is high and the low viscosity fluid forms non-fractal patterns that resemble the fingers of a glove[1,2]. Fractal phenomena are found to be associated with a range of random systems[5–8]. For this reason it is of interest to inquire whether or not there exist experimental conditions under which viscous fingers obey a fractal symmetry.

Experiments

Fractal structures in nature typically occur when randomizing effects dominate over stabilizing effects. Hence to find a fractal structure we must eliminate stabilizing effects so that the randomness can dominate a structure. To this end we shall minimize the role of interfacial tension[1,2] as well as intermolecular diffusion[3]. Therefore, we chose a viscous fluid with zero interfacial tension against water, an aqueous polymer solution of high relative molecular mass polysaccharide for which the mixing time is very slow compared with the time of the experiment.

Since the viscosity difference is the driving force for the instability, a high viscosity is desirable. Hence we use a non-newtonian fluid: the low-shear regions have high viscosity while the shear thinning character of the fluid makes it easy to pump it through the narrow gap of the Hele-Shaw cell without causing

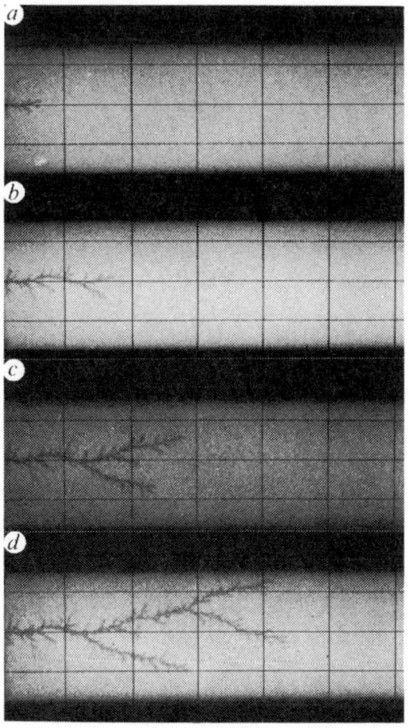

Fig. 1 Typical viscous finger created by water advancing into a Hele-Shaw cell filled with a non-newtonian fluid with zero interfacial tension (experimental conditions described in the text). The time interval between successive photographs is typically 5 s. The grid rectangles have dimensions 3 cm × 5 cm. The water, dyed with methylene blue, is injected under force from a single inlet on the left wall into an aqueous polymer solution whose viscosity varies between 10^3 and 10^4 cP depending on the water velocity.

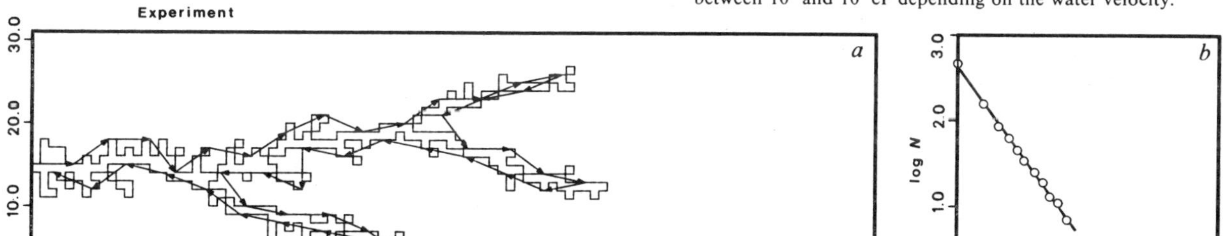

Fig. 2 Schematic illustration of the determination of d_f from experimental data of Fig. 1. *a*, Illustration of how the contour length of the digitized finger is measured with a ruler of length $R = 5$. *b*, The slope of the straight line gives $-d_f$, from which we find $d_f = 1.39$.

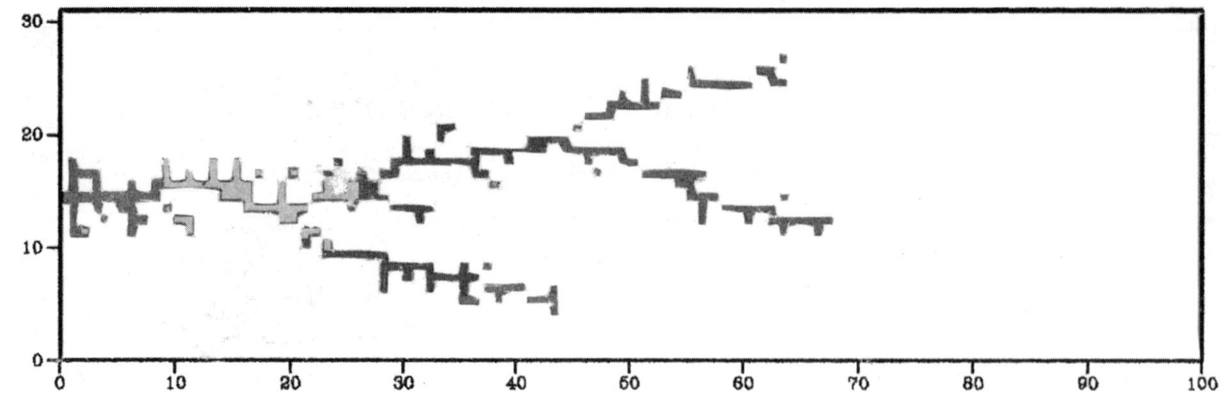

Fig. 3 Digitized representation of the viscous finger of Fig. 1. The colour coding is an indication of the time evolution of the instability, the order being dark blue, light blue, red, violet and pink. The colour coding is for comparison with the model calculations of Fig. 4.

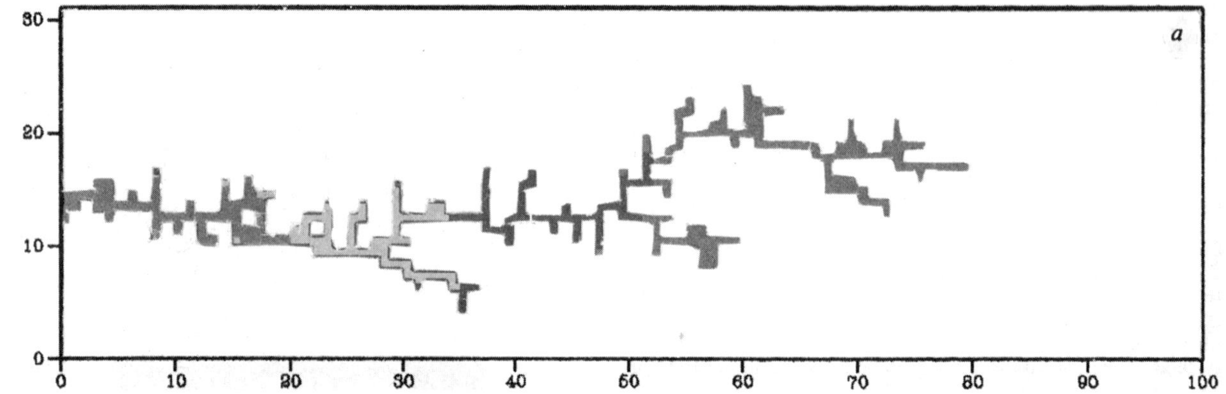

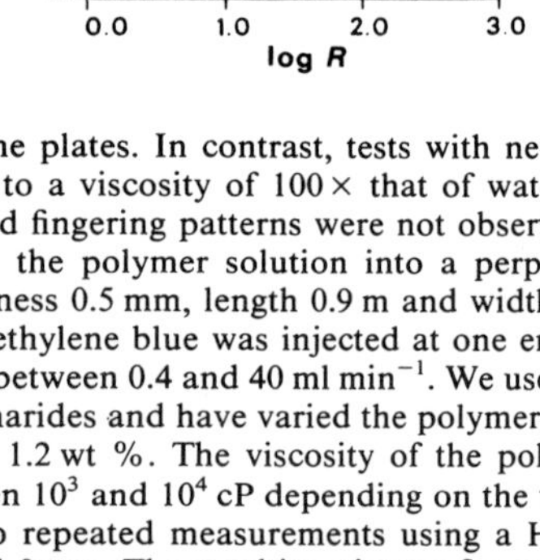

Fig. 4 Results of model calculation for the ratio $W_{HS}/W_f = 30$, corresponding to the experiments of Fig. 1. *a*, Colour coding indicates the time of arrival of the random walker, the order being dark blue, light blue, red, violet and pink; each colour represents 20% of total time. Comparison with Fig. 3 confirms the agreement between experiment and the model. *b*, The fractal dimension of the aggregate shown is 1.39, in agreement with the analysis of the experimental data of Fig. 2.

bending of the plates. In contrast, tests with newtonian fluids were limited to a viscosity of 100× that of water, and steady fine-structured fingering patterns were not observed.

We placed the polymer solution into a perpex Hele-Shaw cell[8] of thickness 0.5 mm, length 0.9 m and width 0.1 m. Water dyed with methylene blue was injected at one end at a variety of flow rates between 0.4 and 40 ml min^{-1}. We used three different polysaccharides and have varied the polymer concentration from 0.24 to 1.2 wt %. The viscosity of the polymer solution varies between 10^3 and 10^4 cP depending on the water velocity. We have also repeated measurements using a Hele-Shaw cell of thickness 1.0 mm. The resulting viscous finger patterns were random and highly branched (Fig. 1, *a–d*).

Analysis

Visual examination of these patterns reveals the striking presence of structure at many different length scales. Hence it seemed appropriate to attempt to analyse the structure of a viscous finger using the concept of self-similarity or fractals[5]. To this end, we first digitized the fingering patterns (Fig. 3) using a pixel size equal to the characteristic thickness of a viscous finger. We then analysed the patterns as follows. We measured the total exterior perimeter length using a 'ruler' of length $R = 1$, measured in units of a pixel. Then we repeated the same operation for a range of increasing R values up to $R = 20$ (Fig. 2*a*). The apparent length of any fractal, in units of R, decreases as R^{-d_f} (Fig. 2*b*)[5,9]. The d_f of the external perimeter is the same as the d_f of the total mass if loops do not occur on all length scales[5]. We find the quantitative parameter, $d_f = 1.40 \pm 0.04$, which characterizes viscous fingers for a wide range of polysaccharides, polymer concentrations and water flow rates, suggesting that viscous fingers indeed have a reproducible and universal fractal dimension.

As a second method, we placed an imaginary box of edge L on each point of the finger. For each box, we counted the number of points (the total mass) inside the box. As L increases, this mass should scale as L^{d_f} if the finger has a fractal structure[10]. We find that Method 2 gives smaller values for d_f than predicted by Method 1. The difference is ~8 % and is explained by the fact that a large proportion of the finger points lie on the

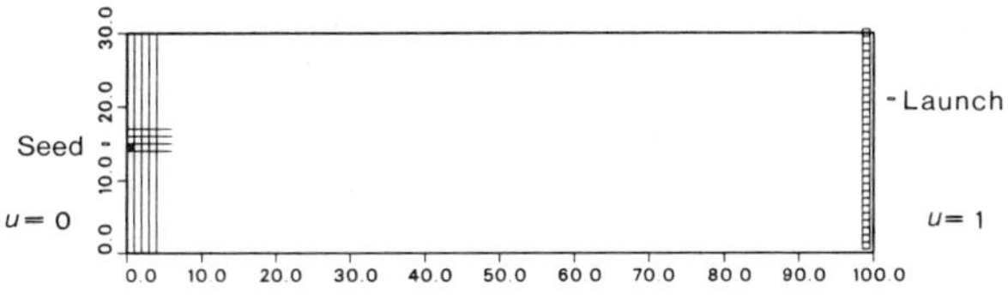

Fig. 5 The basic geometry of the generalized Witten-Sander model used to simulate viscous fingers. Random walkers are released from a randomly chosen point on the right wall, and 'die' if they strike the lateral or end walls.

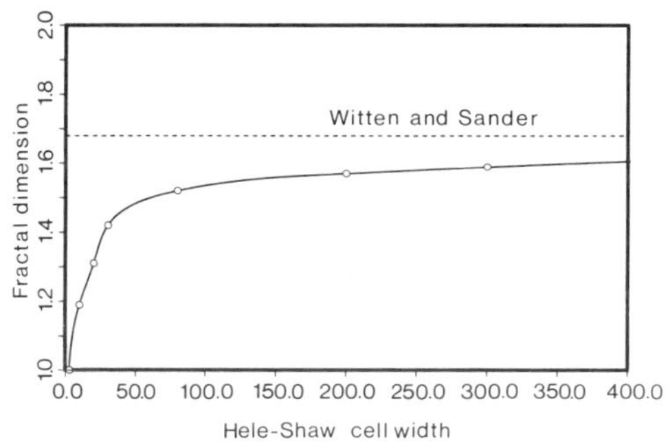

Fig. 6 Dependence on W_{HS} of the apparent value of the fractal dimension, d_f.

perimeter of a particular pattern and boxes centred at these points will cover much unoccupied area outside the structure. For the large-scale numerical simulations discussed below, the number of points on the perimeter is small compared with the total number of points; both methods give identical fractal dimensions.

Statistical model

The mean velocity, v, of a viscous fluid in a Hele-Shaw cell is given by the gradient of a potential $\phi = -(b^2/12\,\mu)\,P$, where b is the plate separation, μ is the shear viscosity and P is the fluid pressure[11]. For incompressible fluids, div $v = 0$; hence we obtain a Laplace equation for the potential distribution

$$\nabla^2 \phi = 0 \qquad (1)$$

If, as in our case, the viscosity ratio between the two fluids is very large, we can approximate the potential in the less viscous fluid to be a constant ($\phi = \phi_0$). At the outlet, ϕ takes on a different value $\phi = \phi_1$.

With these boundary conditions, equation (1) is identical to the Laplace equation describing the probability $u(r, t)$ of a random walker to be at the point r at the time t[12,13]. This is similar to diffusion-limited aggregation (DLA)[13-16], so we propose the following quantitative model for viscous fingering. Place a seed particle at the origin (Fig. 5) and release a random walker from a randomly chosen point on the opposite wall. If the random walker hits the seed, it sticks, whereas if it hits any of the four walls, it 'dies' and a new walker is released. For the simulation, we have chosen our 'Hele-Shaw' cell width $W_{HS} = 30$ (in units of a lattice constant), as the experimental value of $x = W_{HS}/W_f$ is ~30, with W_f the characteristic size of a single finger. Formally, the probability $u(r, t)$ can be mapped onto the fluid potential $\phi(r, t)$ described above, $u(r, t) = (\phi - \phi_0)/(\phi_1 - \phi_0)$. Thus, $\phi - \phi_0$ in the finger corresponds to $u = 0$ in the aggregate, whereas $\phi = \phi_1$ at the outlet maps to $u = 1$ at the source of random walkers.

Two striking features of the actual viscous finger experiments are mirrored by our statistical model. The first of these is the tendency of a growing side branch to die out after a relatively short period of growth. The second is the fact that the entire viscous finger strongly avoids coming even moderately close to the lateral walls of the Hele-Shaw cell. Both effects are seen clearly in the model and arise from a common source, 'screening'[13-16]. This screening can be understood as follows. A random walk, statistically speaking, sticks to the tips of the growing aggregate so that the main trunk of the aggregate effectively shields the smaller side fingers. As the lateral walls absorb the random walkers, they act as 'ghost fingers': the viscous finger and the wall together form an essentially impenetrable deep fjord. This can be seen dramatically by the colour coding of Fig. 4, where the last random walkers to arrive, coloured pink, rarely reach the side fingers and never penetrate the deep fjord between the aggregate and the wall.

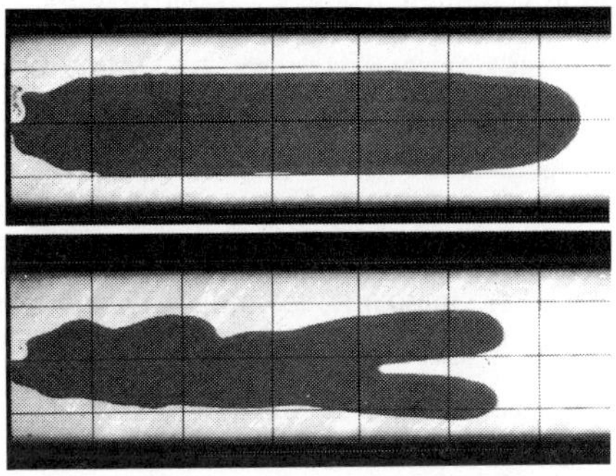

Fig. 7 A viscous finger created by water advancing into a Hele-Shaw cell filled with a newtonian fluid, oil. Two different capillarity numbers are used: *a*, $N_{ca} = 0.08$; *b*, $N_{ca} = 0.16$.

Calculating d_f as described above, we find $d_f = 1.41 \pm 0.05$ (see Fig. 5), a value remarkably close to the experimental value but considerably different from the value $d_f = 1.68 \pm 0.04$ for DLA[17]. Could this difference arise from the geometric constraint of the lateral walls? To test this possibility, we made additional simulations for a range of values of x. We find (Fig. 6) that the apparent value of d_f increases smoothly from $d_f \cong 1$ (for $x \cong 1$) to $d_f \cong 1.60$ (for $x \cong 10^3$). To check this prediction, we also performed experiments for a Hele-Shaw cell half as wide ($x = 15$) and find that $d_f = 1.12 \pm 0.04$, consistent with the predicted value of Fig. 6. What is the source of this apparent dependence of d_f on W_{HS}? At first sight, one might be tempted by an analogy with a growing polymer chain in the same confining geometry; it takes on the bulk value of d_f until it has grown sufficiently to sense the presence of the lateral walls. However, our problem is quite different, since the lateral walls strongly influence even the initial stages of finger growth because the incoming random walks are strongly influenced by the absorbing lateral and end walls. Hence the small fingers cannot capture enough mass to achieve the full DLA structure (this fact becomes especially clear if we consider very small values of W_{HS}). So while for the polymer problem d_f crosses over directly from its bulk value to its ultimate value of $d_f = 1.0$, for the Hele-Shaw fingering problem d_f never takes on its bulk value but instead takes on a smaller value that depends monotonically on W_{HS} (Fig. 6), crossing over ultimately to $d_f = 1$ when $R \gg W_{HS}$.

Discussion

We consider here the relation of the present experiments to previous work on viscous fingering. The classic Saffman-Taylor work did not produce ramified fractal structures. To obtain fine-structured instability patterns for an immiscible fluid system, the critical wavelength λ_c of a perturbation[2,18], defined as $\lambda_c = k[b^2/(12\,N_{ca})]^{1/2}$ has to be much smaller than W_{HS}. N_{ca} is the capillarity number, defined as $N_{ca} = \mu v/\sigma$ (σ is the interfacial tension) and corresponding to the ratio of viscous forces to capillarity forces: k is a dimensionless parameter that depends on the contact angle between the two fluids and the plates (wettability)[18]. We have also performed Saffman-Taylor type experiments with an immiscible newtonian oil-water system with varying flow velocity v, and Fig. 7 indicates that a change in the flow condition from $N_{ca} = 0.08$ to 0.16 introduces a bifurcation into fingers when the finger thickness exceeds twice λ_c[19]. A continuation of this process would presumably create smaller and smaller fingers with a tree-like structure, but would require a further increase in the capillarity number or a decrease of the plate distance b. We cannot test the possibility that newtonian fluids with very low interfacial tension and negligible intermolecular diffusion but with a very high viscosity ratio, might give rise to similar fractal instability patterns, as our experiments were limited by the finite strength of the cell. By using water and a non-newtonian water-based fluid, we obtained a flow with an infinite capillarity number and a high viscosity ratio in regions where shear stresses are low. We believe both aspects are important and explain the difference from the Saffman–Taylor results.

Our present findings can be applied to the basic physics underlying petroleum science[4,20]. In the traditional '5-spot well configuration', four production wells are drilled at the corners of a large square and CO_2 (or a low viscosity fluid with a very low interfacial tension with oil) is then pumped into a fifth injection well in the centre. The CO_2 plays the role of the water in our model experiments, whereas the oil is analogous to the polymer solution. Large-scale viscous fingering phenomena not bounded by the walls of a Hele-Shaw cell occur and drastically reduce the efficiency of the oil recovery. The observed viscous fingers are quasi-two-dimensional as the thickness of the rock formation is typically several orders of magnitude less than the separation of the injection and production wells. It is therefore customary to assume that the patterns observed in the quasi-two-dimensional Hele-Shaw cells are relevant to the observed macroscopic phenomena. The similarity in the mathematical description of flow through porous media and the Hele-Shaw cell have been observed previously[1]. Thus, our experiments and model calculations strongly suggest that the viscous fingers observed in a 5-spot pattern are rich geometrical structures possessing a scale-invariant symmetry and quantitatively characterized by a fractal dimensionality.

Conclusion

There is no complete theory of miscible viscous fingering, perhaps because there was no quantitative parameter to characterize viscous fingers. Here, we find that viscous fingers are fractals and that their many different random structures have the identical value of the fractal dimensionality. This suggests that the physical basis of the viscous fingering instability is analogous to that of other second-order phase transitions that obey the same sort of scale invariant symmetry found here for viscous fingers[21]. Recent work supports our point of view: quasi-two-dimensional dielectric breakdown seems to exhibit the same sort of fractal structure that we find for the viscous fingering instability[22]. That these two breakdown phenomena should be related is plausible in light of the parallels in the equations describing them[23].

We thank R. Lemanczyk, P. Rossi and the management of Dowell Schlumberger for their encouragement during this work. The Center for Polymer Studies is supported by grants from NSF and ONR. We also thank F. Leyvraz, P. Meakin and J. D. Sherwood for helpful discussions.

Note added in proof: Since acceptance of this article, we have learned of independent work on viscous fingering, both from the experimental (ref. 24, and L. M. Sander and E. Ben-Jacob, personal communication) and theoretical[25,26] points of view.

Received 31 October 1984; accepted 24 January 1985.

1. Saffman, P. G. & Taylor, G. I. *Proc. R. Soc.* **A245,** 312–329 (1958).
2. Chuoke, R. L., Van Meurs, P. & Van der Poel, C. *J. petrol. Tech.* **11,** 64–70 (1959).
3. Stalkup, F. I. *Miscible Displacements* (Soc. of Petroleum Eng., AIME, N.Y., 1983).
4. Hele Shaw, J. S. S. *Nature* **58,** 34–36 (1898).
5. Mandelbrot, B. B. *The Fractal Geometry of Nature* (Freeman, San Francisco, 1982).
6. Brady, R. M. & Ball, R. C. *Nature* **309,** 225–229 (1984).
7. Avnir, D., Farin, D. & Pfeiffer, P. *Nature* **308,** 261–263 (1984).
8. Matsushita, M., Sano, M., Hayakawa, Y., Honjo, H. & Sawada, Y. *Phys. Rev. Lett.* **53,** 286–289 (1984).
9. Powles, J. G. & Quirke, N. *Phys. Rev. Lett.* **52,** 1571–1574 (1984).
10. Forrest, S. L. & Witten, T. A. *J. Phys.* **A12,** L109–112 (1979).
11. Lamb, H. *Hydrodynamics* (Cambridge University Press, London, 1932).
12. Chandrasekhar, S. *Rev. mod. Phys.* **15,** 1–89 (1943).
13. Witten, T. A. & Sander, L. M. *Phys. Rev. Lett.* **47,** 1499–1501 (1981).
14. Witten, T. A. & Sander, L. M. *Phys. Rev.* **B27,** 5685–5697 (1983).
15. Herrmann, H. J. *Phys. Rep.* (in the press).
16. Stanley, H. E. *Fractals in Statistical Physics* (Oxford University Press, New York, 1985).
17. Meakin, P. *Phys. Rev.* **A27,** 1495–1507 (1983).
18. Peters, E. J. & Flock, D. L. *J. Soc. petrol. Engng* **21,** 249–258 (1981).
19. Paterson, L. *J. Fluid Mech.* **113,** 513–529 (1981).
20. Todd, M. R. & Longstaff, W. J. *J. petrol. Tech.* **253,** 874–882 (1972).
21. Stanley, H. E. *Introduction to Phase Transitions and Critical Phenomena* (Oxford University Press, 1971).
22. Niemeyer, L., Pietronero, L. & Wiesmann, H. J. *Phys. Rev. Lett.* **52,** 1033–1036 (1984).
23. Paterson, L. *Phys. Rev. Lett.* **52,** 1621–1624 (1984).
24. Maher, J. V. *Phys. Rev. Lett.* (submitted).
25. Tang, C. *Phys. Rev. A* (submitted).
26. Kadanoff, L. P. *J. stat. Phys.* (submitted).

Printed in Great Britain by Macmillan Production Ltd., Basingstoke, Hampshire

Reprinted with permission from the *Material Resources Society Bulletin* **12**, J.P. Gollub and L.M. Sander, "Pattern Formation in Materials Science," pp. 98–100, ©1987.

Pattern Formation in Materials Science

J.P. Gollub and L.M. Sander

Introduction

The growth of materials at interfaces frequently leads to patterns with length scales that can be much larger than atomic sizes. Patterns resulting from morphological instabilities occur during crystal growth and are responsible for the intricate shapes of snowflakes, dendritic crystals, and the like. They also occur during the electrochemical deposition of metals on surfaces, and during the growth of thin films by vapor deposition. Our purpose is to point out some particularly interesting connections that have come to be appreciated during the past five years between different types of pattern-forming phenomena, and to summarize some recent theoretical approaches to understanding them.

Crystal Growth

The phenomenon of dendritic crystal growth, though studied for many years, is still a great challenge. As an example, we show the development of a needle crystal of ammonium bromide (NH_4Br) from supersaturated aqueous solution in Figure 1. The contours are cross sections of the interface at 20-second intervals, obtained by digital image analysis. One can see that an approximately parabolic tip translates at constant speed, and that the needle crystal is apparently unstable to the development of a train of sidebranches that propagate outward from the main stem but do not move forward with the tip.

Some of the basic questions about this structure are: How fast does it grow? Why is the tip nearly parabolic, and what determines its radius of curvature? What causes the sidebranches, and how is their spacing determined? Where do the sidebranches start? Why is the pattern irregular? Finally, what are the fundamental requirements for dendritic growth?

Much literature covers these questions, and significant progress has been made the past few years by analysis of a model in which the transport is assumed to be governed by molecular or heat diffusion in front of the interface, with boundary conditions that include two essential features — surface tension and crystalline anisotropy.[1-4] The hexagonal structure of ice crystals was noted as early as the 2nd Century B.C., as noted in a historical review by Hobbs.[5] However, the actual incorporation of anisotropy into a quantitative theory is quite recent.

These considerations yield a well-defined model system that has been studied mathematically in considerable depth. Although the diffusion equation itself is linear, the moving boundary makes the overall problem highly nonlinear and hence difficult to solve. In a two-dimensional analysis of the equations, there appears to be a unique stable solution having the form of an unbranched needle crystal. Other solutions exist but are believed to be unstable to perturbations that change the tip curvature. Predictions have been made for the tip curvature and velocity for several materials, and comparisons with experiments are in progress. Some inconsistencies have been noted (e.g., with experiments on succinonitrile[6]), but the comparisons are not yet complete. It is possible that equilibrium measurements of the crystalline anisotropy, which are made by examining the shape of a small crystallite in equilibrium, may prove insufficient for a proper comparison between theory and experiment. Additional anisotropy due to kinetic effects may be manifested during nonequilibrium growth.

If it is true that noise is important in determining the structure of a dendrite, then the limitations of a purely continuum model of the process become apparent.

The problem of understanding the intricate sidebranching behavior[6] is even more difficult. It is apparent that the tips are relatively stable, but that the sides of the dendrite are not. One recent view is that a finite amplitude instability is stimulated by "noise" near the tip, which might possibly be microscopic noise due to concentration or temperature fluctuations.[7] Careful recent measurements[8] of the structure near the tip of the dendrite shown in Figure 1 do in fact reveal that the sidebranches begin very close to the tip and are irregular even when they are small (less than one micron), lending some support to this hypothesis.

If it is true that noise is important in determining the structure of a dendrite, then the limitations of a purely continuum model of the process become apparent. This, and a tentative understanding of the critical role of surface tension and anisotropy, are significant new theoretical insights that still await experimental verification.

Theoretical work on morphological problems relevant to crystal growth has been enriched by an appreciation that pattern formation in certain fluid systems may be analogous. Consider, for example, a thin layer of fluid that is displaced by injection of a less viscous (and immiscible) second fluid. The important dynamical variable for this problem is the pressure field, which is governed by Laplace's equation if the layer is thin enough.

The reason that fluid displacement and crystallization may be related is that Laplace's equation and the diffusion equation are the same for steady state processes. Therefore both show interfacial instabilities. In fluid displacement the patterns are generally more irregular.[9] However, if anisotropy is added to the fluid system, dendritic patterns can be realized. This can be done either by etching grooves[10] on the confining plates as in Figure 2, or by placing a small bubble at the tip

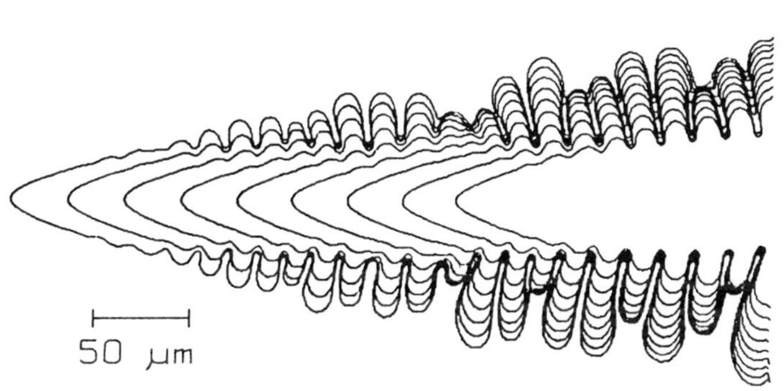

Figure 1. Contours of an NH_4Br dendrite growing from supersaturated solution, at 20-second intervals. From Reference 8 with permission.

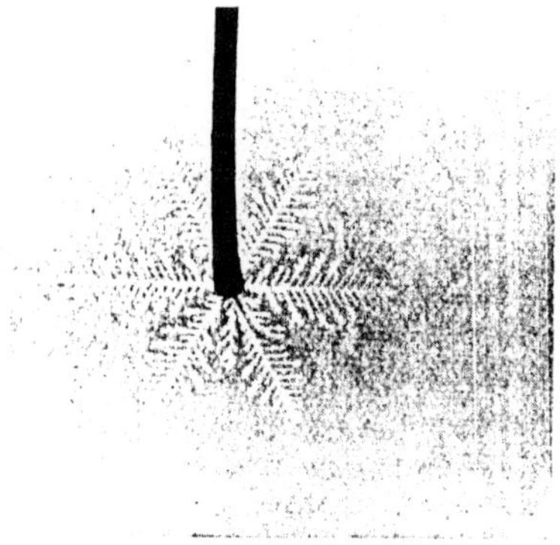

Figure 2. Dendritic growth during immiscible fluid displacement, after Reference 10.

of the advancing fluid.[11] These experimental observations show very graphically that anisotropy is an essential requirement for dendritic structures.

Before leaving this discussion of crystal growth, we point out that the electrochemical deposition of metals also yields a variety of morphological phenomena.[12,13] Both dendrites and finely divided structures that are well approximated as fractals can be realized by varying the applied electrical potential between the center and periphery of a thin circular growth cell filled with an electrolytic solution of a zinc salt (Figure 3). The microstructure of the crystals thus formed seems to correlate with their growth habit: dendritic deposits have large crystallites and the fractals very small ones (about 50 Å).

Film Growth by Vapor Deposition

In the formation of a dendritic crystal the mean free path of the attaching molecules is very much shorter than the size of the growing crystal. The crystal is nearly in local equilibrium, though the overall pattern is determined by kinetic effects. It is interesting to contrast this with the situation in which the molecules have a long mean free path, as might be the case for film growth by vapor deposition. Here the mean free path of the accreting molecules can be hundreds of times longer than the thickness of the film.[14] An interesting idealization of this process is the ballistic deposition model,[15] where particles are assumed to deposit irreversibly after being projected in parallel straight lines. Though this model seems much simpler than that for dendritic growth, theoretical investigation is just beginning[16,17] and many questions remain open.

The fundamental instability which gives rise to needle crystals is absent in this case; uniform amorphous films can be grown by vapor deposition. However, a noise-driven nonlinear instability seems to occur for non-normal incidence; it gives rise to the columnar microstructure which is illustrated in a computer simulation in Figure 4. Effects very much like the computer picture seem to occur in the deposition of real materials.[14] The long, irregular columns are seen in many variations of the model and in a variety of experiments. Curiously enough, the columns do not line up with the incident beam, but are tilted toward the normal. A partly successful analytic theory given for the column formation involves both the tilting of the columns and the shadowing effect between columns.[18]

Disorderly Crystallization and Fractals

In some cases the complex but orderly patterns of dendritic crystallization give way to sprawling, irregular structures such as those in Figure 3b; they seem to lack any definite overall pattern. In fact, there is still a symmetry in these cases, but of a subtle kind, namely fractal scaling.[19] A fractal is a self-similar object: it is an image of itself when viewed on a different scale. For the randomized crystals that concern us here, this self-similarity is true only in a statistical sense.

The possibility that objects which grow by diffusion might be fractal was pointed out by Witten and Sander,[20] who devised the diffusion-limited aggregation model (DLA). DLA may be regarded as a discrete version of the dendritic crystallization theory mentioned above (in the Laplace limit), but without surface tension or anisotropy and with a good deal of intrinsic noise. The computer simulation pictures made with the model look almost exactly like the pattern in Figure 3b, and both show well-defined self-similarity.

Other sorts of crystallization, such as the annealing of surface films of $GeSe_2$,[21] have been shown to give rise to DLA-type fractals. In fact, DLA patterns have been found in a variety of contexts: for example, the patterns of dielectric breakdown channels (known as Lichtenberg figures) have turned out to be fractals of exactly the same sort.[22]

The relationship of fractal crystals to ordinary dendrites is under scrutiny. Much remains to be learned about the transition between the two regimes, but it now appears that when external noise overwhelms the effects of anisotropy, the stable tip of the dendrite splits and the proliferation of these tip-splittings gives rise to a fractal.[23]

It is also interesting to contrast DLA to ballistic aggregation. In DLA the accreting particles diffuse, i.e., follow random walk trajectories, rather than the straight-line paths of ballistic aggregation. This change is enough to convert a sprawling, open fractal or a needle crystal into a solid film. The only remaining vestige of scaling is in

...it now appears that when external noise overwhelms the effects of anisotropy, the stable tip of the dendrite splits and the proliferation of these tip-splittings gives rise to a fractal.

the roughness of the upper surface of the film.[16,17]

Summary

We have discussed three examples of patterns formed during material growth. A theory of the dendritic needle crystal, for example, seems to be in hand, though not yet confirmed experimentally. The extension of the theory to explain initial side-

continued

Figure 3. Dendritic and disordered growth during the electrolytic deposition of zinc from water solution.

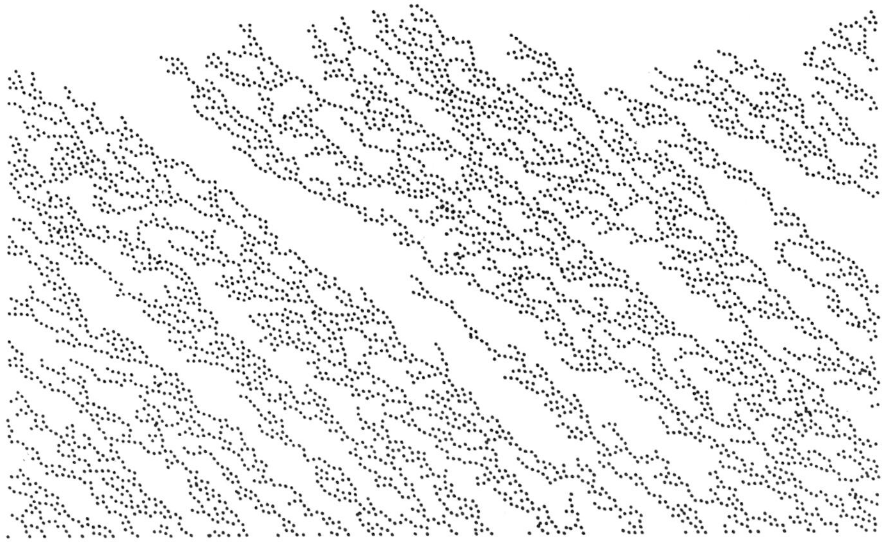

Figure 4. Computer simulation of ballistic aggregation showing the columnar microstructure.

branching may be possible, but the fully developed structure is still out of reach. Finally, we are just beginning to find a framework to organize our ideas about ballistic aggregation and random fractal crystallization.

Acknowledgments

J.P. Gollub wishes to acknowledge the support of grant DMR-8503543 from the NSF Low Temperature Physics Program. L.M. Sander is supported in part by grant DMR-8505474 from the NSF Condensed Matter Theory program.

References

1. D.A. Kessler, J. Koplik, and H. Levine, *Phys. Rev. A* **33** (1986) p. 2621 and 2634; and *Advances in Physics*, to be published.
2. J.S. Langer, Les Houches Lectures (1986), to be published.
3. D. Meiron, *Phys. Rev. A* **33** (1986) p. 2704.
4. Y. Saito, G. Goldbeck-Wood, and H. Muller-Krumbhaar, *Phys. Rev. Lett.* **58**, (1987) p. 1541.
5. P.V. Hobbs, *Ice Physics* (Clarendon Press, Oxford, 1974), p. 524.
6. S-C. Huang and M.E. Glicksman, *Acta Met.* **29** (1981) p. 701 and 717.
7. R. Pieters and J.S. Langer, *Phys. Rev. Lett.* **56**, (1986) p. 1948; D.A. Kessler and H. Levine, *Phys. Rev. Lett.* **57** (1986) p. 3069; B. Caroli, C. Caroli, and B. Roulet, private communication.
8. A. Dougherty, P.D. Kaplan, and J.P. Gollub, *Phys. Rev. Lett.* **58** (1987) p. 1652.
9. D. Bensimon, L.P. Kadanoff, S. Liang, B.I. Shraiman, and C. Tang, *Revs. Mod. Phys.* **58** (1986) p. 977; G.M. Homsy, *Ann. Rev. Fluid Mech.* **19** (1987) p. 271.
10. E. Ben-Jacob, R. Godby, N.D. Goldenfeld, J. Koplik, H. Levine, T. Mueller, and L.M. Sander, *Phys. Rev. Lett.* **55** (1985) p. 1315.
11. Y. Couder, N. Gerard, and M. Rabaud, *Phys. Rev. A* **34**, (1986) p. 5175.
12. Y. Sawada, A. Dougherty, and J.P. Gollub, *Phys. Rev. Lett.* **56** (1986) p. 1260.
13. D. Grier, E. Ben-Jacob, R. Clarke, and L.M. Sander, *Phys. Rev. Lett.* **56** (1986) p. 1264.
14. H.J. Leamy, G.H. Gilmer, and A.G. Dirks, in *Current Topics in Materials Science*, edited by E. Katdis (North-Holland, 1980).
15. M.J. Vold, *J. Colloid Interface Sci.* **18**, (1963) p. 684.
16. M. Kardar, G. Parisi, and Y-C. Zhang, *Phys. Rev. Lett.* **56** (1986) p. 889.
17. P. Meakin, P. Ramanlal, L.M. Sander, and R.C. Ball, *Phys. Rev. A* **34** (1986) p. 5091.
18. P. Ramanlal and L.M. Sander, *Phys. Rev. Lett.* **54** (1985) p. 1828.
19. B. Mandelbrot, *The Fractal Geometry of Nature* (Freeman, San Francisco, 1982).
20. T. Witten and L.M. Sander, *Phys. Rev. Lett.* **47** (1981) p. 1400; *Phys. Rev. B* **27** (1983) p. 1119.
21. G. Radnoczi, T. Viscek, L.M. Sander, and D. Grier, *Phys. Rev. A* **35** (1987) p. 4012.
22. L. Niemeyer, L. Pietronero, and H. Wiesmann, *Phys. Rev. Lett.* **52** (1984) p. 1033.
23. H. Honjo, S. Ohta, and M. Matsushita, *J. Phys. Soc. Japan* **55** (1986) p. 2487.

Jerry P. Gollub is professor of physics at Haverford College and adjunct professor at the University of Pennsylvania. His research concerns pattern formation during solidification, hydrodynamic instabilities, and nonlinear dynamics.

Leonard M. Sander is professor of physics at the University of Michigan. His research concerns theories of nonequilibrium growth, aggregation, and fractals.

Fractal growth of copper electrodeposits

R. M. Brady & R. C. Ball

Cavendish Laboratory, University of Cambridge, Madingley Road, Cambridge CB3 0HE, UK

Black dendritic growths of copper were electrodeposited in steady-state diffusion-limited conditions onto an initially point-like cathode. Their spatial extent R *was inferred from the increased current and their mass* M *from the total charge transferred. Fractal behaviour* $M \propto R^D$ *was observed with* $D = 2.43 \pm 0.03$, *in good agreement with computer simulations of diffusion-limited aggregation.*

IF the accretion of material onto a surface is limited by diffusion onto the surface, deposition occurs preferentially on any protuberances (just as electric field lines concentrate on the convexities of a lightning conductor). In this way any initial roughness grows exponentially—the Mullins–Sekerka (MS) instability—and the deposit rapidly acquires a dendritic quality. Surface properties limit how finely ramified the deposit grows. This instability has been long recognized by materials scientists, with examples ranging from alloy solidification patterns to dendritic crystal growth (see ref. 1 for a review).

Recently, Witten and Sander[2] introduced a simple computer model in which particles diffuse onto a cluster by random walk, one at a time. This model differs from experimental systems principally by the absence of any surface tension or crystalline factors and by the very slow rate of reaction; and in this 'diffusion-limited aggregation' (DLA) model the clusters grow immediately dendritic and disorderly. They were shown to be fractal[3] in the sense that inside a large cluster the number of particles within a radius R of any given particle was on average proportional to R^D. The index D is the Hausdorff dimensionality[4] of the deposit and was found by computer simulation in three dimensions by Meakin[5] to be $D = 2.495 \pm 0.06$. Note that this model does not seem to be applicable to experiments (ref. 6 and D. Weitz, in preparation) where clustering of clusters was allowed and $D = 1.7$ obtained.

Even in the experimental systems where the MS instability is

226 ARTICLES NATURE VOL. 309 17 MAY 1984

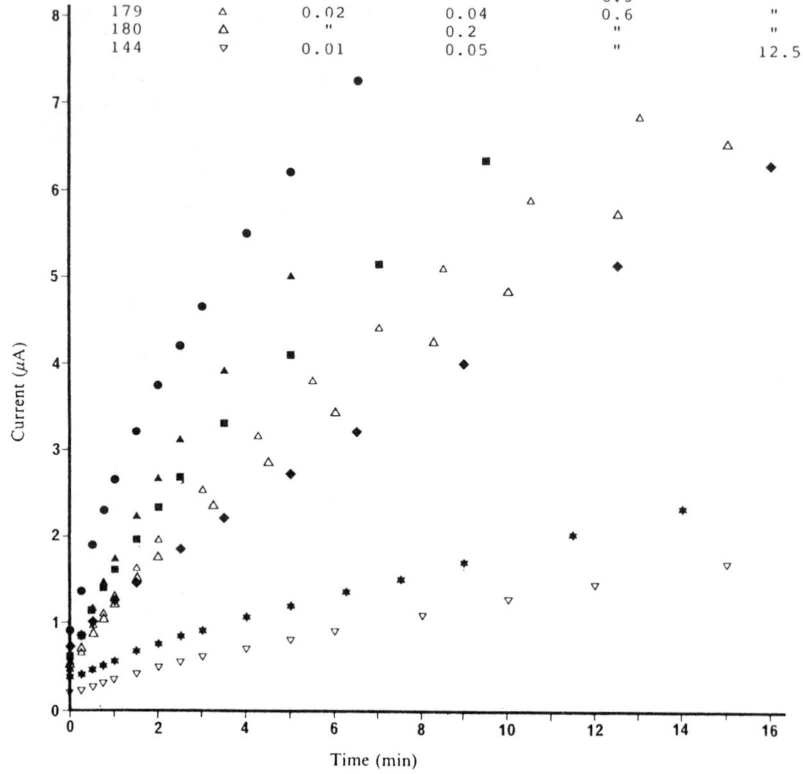

Run no.	Symbol	$[CuSO_4]$ (M)	$[Na_2SO_4]$ (M)	Potential (V)	Wire diameter (μm)	Temperature (°C)
173	●	0.029	0.05	0.8	25	22
163	▲	"	"	0.6	"	23
174	■	"	"	0.45	"	23
176	◆	"	"	0.375	"	23
172	✱	"	"	0.3	"	22
179	△	0.02	0.04	0.6	"	26
180	△	"	0.2	"	"	26
144	▽	0.01	0.05	"	12.5	24

Fig. 1 Increase of current with time, for various potentials and electrolyte concentrations. All the solutions contained 1.4 wt% polyethylene oxide to impede convection.

limited by surface properties, Witten and Sander[2] suggested that weak noise should ultimately lead to disorder and similar fractal behaviour. Thus, at short length scales there may be local order due to surface tension or crystallographic factors, but at longer length scales they suggested random fluctuations should dominate: in particular, the Hausdorff dimensionality D found by computer simulation should also characterize experimental systems at long length scales.

We now report an experimental measurement of the Hausdorff dimensionality D of copper deposited electrolytically from aqueous solution in diffusion-limited conditions, onto an initially point-like cathode. The deposits were grown larger than the initial cathode but smaller than any diffusion length in the system. Because the diffusion-limited current onto such a spherical absorber is proportional to its radius (and[7] not its surface area), the instantaneous current gave an effective overall radius R_s for the deposit. With the mass M of the deposit determined from the total charge transferred, the relationship between M and R_s could then be inferred, and the fractal law $M \propto R_s^D$ verified.

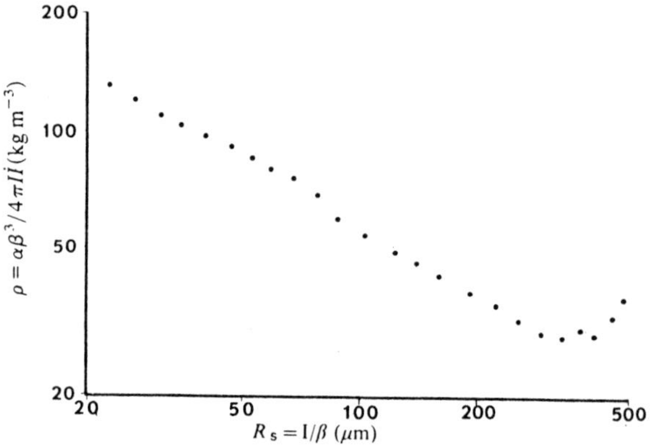

Fig. 2 Logarithmic plot of inferred density as a function of radius, for run 144.

Experiment

Solutions of copper sulphate and sodium sulphate were made up using deionized, distilled water. Polyethylene oxide $(C_2H_4O)_n$ with molecular weight 6×10^6 was added at 1.4 wt%, increasing the viscosity by a factor of 1,000 to ~1 P. The solutions were stored away from light and vigorous stirring was avoided, to prevent damage to the polymer.

In each experiment a sample of the solution was placed inside a hole of diameter 7 mm drilled in a copper block, which acted as the anode. Insulated wire of diameter 12.5 or 25 μm, the end freshly cut with a razor blade, acted as the cathode. The copper deposit would grow in a dendritic cluster around the cut end[4] of the wire which was suspended vertically into the solution. To prevent mechanical movement during the experiment the free length of wire was kept to <7 mm, a vibration-free mounting covered with a bell-jar was used, and at least 5 min were allowed for any movement to die down before the voltage was applied.

The experiments were performed at constant voltage set in the range 0.3–0.8 V, and the current through the cell was measured as a function of time. Initially the interval between readings was 10 s, increasing to several minutes as the current rose more slowly. Typically the experiments were stopped after half an hour.

NATURE VOL. 309 17 MAY 1984

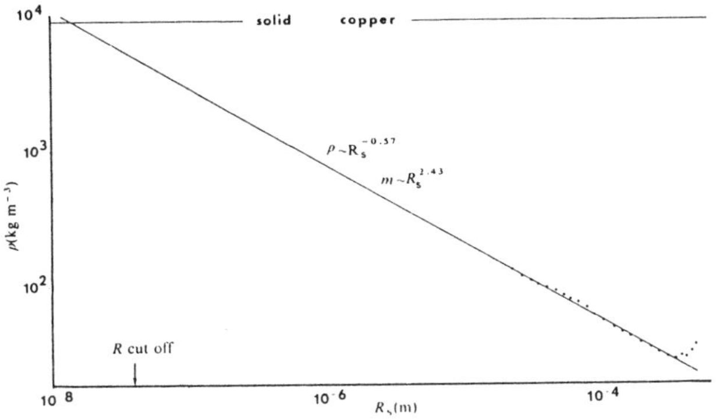

Fig. 3 The data of Fig. 2 plotted on a scale showing the density of solid copper and the radius $R_{\rm cutoff}$ of crystallites observed in electron micrographs. Note the extrapolation (solid line) corresponds to a Hausdorff dimensionality of 2.43.

Figure 1 shows direct plots of current I against time t for various voltages and electrolyte concentrations. It can be seen that the current grows much faster for higher voltages and higher concentrations of copper sulphate, but is relatively insensitive to the sodium sulphate concentration.

It was noted that changing the instantaneous voltage did not change the instantaneous current significantly in our experiments. Any change would have indicated that the reaction was not limited by diffusion of Cu^{2+} ions but was dominated by some other mechanism. The set voltage thus influenced the current–time curves through the morphology of the deposit and not directly through any ohmic or activation limit on the conduction.

Figure 2 shows, with interpretations discussed below, the rate of change of current dI/dt plotted as $-\log I(dI/dt)$ against $\log I$. Plotted in this way the experiment was reproducible to better than 10% except at high currents. Problems at high current and hence large radii included mechanical motion (for example, the wire bending under the weight of the deposit), stirring of the solution, and contact resistances (later improved by applying silver dag).

Results and interpretation

We define an effective radius for the deposits in terms of the diffusion-limited current. Modelling the cathode growth by a perfect spherical absorber of radius R_s, the current[7] is

$$I = \beta R_s \tag{1}$$

the coefficient β being given by

$$\beta = 4\pi K C q \tag{2}$$

where K is the diffusion coefficient, q the charge, and C the number density far from the absorber of the copper ions. This 'Smoluchovski radius' R_s characterizes where the deposition is taking place, and thus how far out the deposit extends in space. Clearly R_s will be of the same order of magnitude as the radius of gyration, R_g, measured in computer simulations[2,5]. Indeed, provided the growths are self-similar fractals as the simulations suggest, within the diffusion-limited regime the ratio R_s/R_g will be a constant independent of all parameters.

The number N of copper atoms deposited can be inferred from the charge transferred Q,

$$N = Q/q = \alpha \int I \, dt \tag{3}$$

Using equations (1)–(3), the relationship between the radius R_s and the number of atoms N can in principle be observed directly throughout the growth of the deposit. In practice the cathode was initially of finite size so that there would be an offset error in the N obtained by integration. It is to avoid this that we have instead plotted dI/dt in Fig. 2.

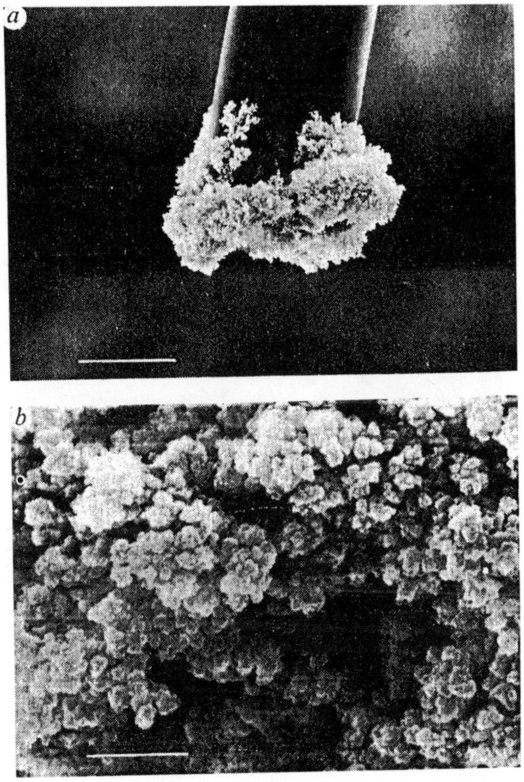

Fig. 4 Electron micrograph of a deposit grown for only 1 min in the absence of polymer. 0.01 M $CuSO_4$, 0.05 M Na_2SO_4, 0.6 V. Scale bars: *a*, 20 μm; *b*, 1 μm.

The quantity

$$\rho = \frac{\beta^3 \alpha}{4\pi I \frac{dI}{dt}} \tag{4}$$

may be interpreted as an effective density of deposition since, using the definition (1) above,

$$dN = 4\pi R_s^2 \, dR_s \rho \tag{5}$$

Figures 2 and 3 show the density–radius plot inferred from the data shown in Fig. 1. Note the logarithmic straight line $\rho \propto R^{-0.57}$ which implies fractal behaviour, $N \propto R^D$ with $D = 2.43$. Writing this result in the form

$$\rho = \rho_{Cu}(R/a)^{D-3} \tag{6}$$

where ρ_{Cu} is the density of solid copper, the length a characterizes the onset of fractal behaviour. Extrapolating the data in Fig. 3 back to the solid copper density gives an intercept of

$$a = 14 \text{ nm} \tag{7}$$

which compares with the characteristic radius of the dendritic features seen in the electron micrographs (Fig. 4)

$$R_{\rm cutoff} \simeq 40 \text{ nm} \tag{8}$$

The specimen in the micrographs was produced in similar conditions, except that it was grown for only 1 min and with no polymer present so as to minimize damage on removal from the solution.

The experiment was repeated at various electrolyte concentrations as listed in Table 1 together with the Hausdorff dimensionalities D obtained. The values of D show no evidence of systematic variation and taken together give

$$D_{\text{experiment}} = 2.43 \pm 0.03 \quad (9)$$

which compares with the result from DLA computer simulations[5]

$$D_{\text{DLA simulation}} = 2.495 \pm 0.06 \quad (10)$$

For each run the intercept a was extrapolated using the value $D = 2.43$, and Fig. 5 shows how the reciprocal radius $1/a$ varied with the voltage of the experiment. We have no explanation for the apparent law,

$$1/a \propto V - (0.23 \text{ V}) \quad (11)$$

Note that the intercept a in Fig. 5 does not depend on the concentration of copper sulphate.

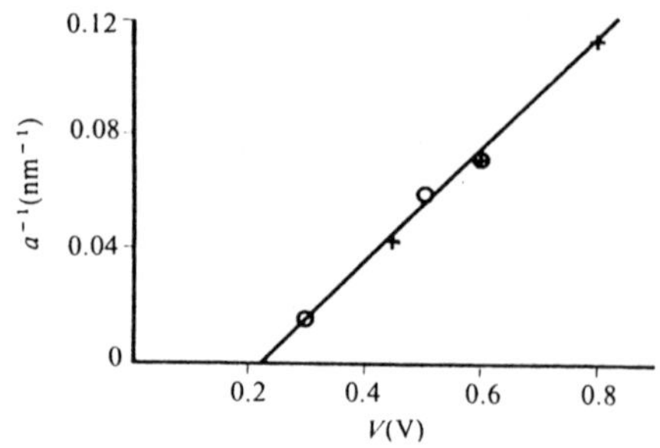

Fig. 5 Variation with voltage of intercept radius a. 0.05 M Na_2SO_4, 1.4 wt% polyethylene oxide. Concentration of copper sulphate (redetermined by solution conductivity): +, 0.029 M; ○, 0.01 M.

Table 1 Fractal dimension of copper electrodeposited from solution at 0.6 V

Run no.	[$CuSO_4$] (M)	[Na_2SO_4] (M)	Measured dimension
144	0.01	0.05	2.43
146	0.01	0.05	2.425
150	0.01	0.05	2.465
153	0.01	0.05	2.485
177	0.02	0.04	2.38
178	0.02	0.2	2.44
179	0.02	0.04	2.425
180	0.02	0.2	2.41
Mean dimension			2.433 ± 0.03
Computer prediction[5]			2.495 ± 0.06

The solutions contained 1.4 wt% polyethylene oxide to increase viscosity.

Electrochemical considerations

Our interpretation of the experiment presupposes that the net reaction at each electrode was

$$Cu^{2+}(aq) + 2e^- \rightleftharpoons Cu(s) \quad (12)$$

and that the reaction was limited by diffusion of Cu^{2+} ions in the vicinity of the cathode.

To evolve hydrogen gas at the cathode in the principal experimental conditions (0.01 M $CuSO_4$, 0.05 M Na_2SO_4, neutral pH) requires at least 0.685 V (ref. 8). Below this limit there are no anion oxidations available, and the only remaining chemical difficulty is the role of singly charged Cu^+ ions. In experimental conditions there is an equilibrium ratio in the presence of copper (see ref. 8),

$$[Cu^+]/[Cu^{2+}] = 7 \times 10^{-3} \, (25\,°C, 0.01 M\, Cu^{2+}) \quad (13)$$

and such a concentration of Cu^+ is soluble. With the anode working at low current density (of the order of 1 $\mu A\, cm^{-2}$), we presume that this equilibrium was maintained and therefore that the current due to Cu^+ ions was negligible.

If stirring of the solution is prevented, the electric fields will be screened and deposition will be limited by diffusion in the neighbourhood of the cathode. The calculated Debye–Huckel screening length for 0.05 M sodium sulphate is $\lambda = 1$ nm. (There is in fact a small unscreened voltage of the order of

$$V = kT/e([Cu^{2+}]_{\text{bulk}} - [Cu^{2+}])/[Na_2SO_4] = 5 \text{ mV} \quad (14)$$

required to maintain neutral charge density as the Cu^{2+} ions adopt a density profile governed by diffusion kinetics.)

Data from the Smithsonian Tables[9] indicate an ionic diffusion constant for aqueous Cu^{2+} at 24 °C of $K = 6.17 \times 10^{-10}$ m^2 s^{-1}, with a temperature coefficient of 0.0229 K^{-1}. This implies a value for our parameter β (equation (1)) of

$$\beta = 0.0121 \quad \text{A m}^{-1} (\pm 5\%) \quad (15)$$

We assume that the polymer had little effect so that the main error in the value for β was the use of dilute solution approximations.

Diffusion lengths

The assumption of Smoluchovski kinetics (equations (1), (2)) requires that the cathode radius R is less than any intrinsic diffusion lengths limiting the scope of a laplace equation for [Cu^{2+}].

The rate at which the diffusant concentration can relax determines a diffusion length $\xi_t = \sqrt{Kt}$. From equations (1)–(4) the dimensionless ratio $\gamma_t = R/\xi_t$ is of the order of

$$\gamma_t = \sqrt{C/\rho} \quad (16)$$

reaching a maximum value of 0.1 for the largest deposits grown (see Fig. 3) with a concentration of copper suphate $C = 0.01$M.

The fluid close to the cathode is depleted of copper ions and is thus less dense by about 2% than the surrounding fluid in experimental conditions. Comparing the resulting flux due to convection with that due to diffusion gives a dimensionless parameter

$$\gamma_c = R/\xi_c = mgCR^3/\eta K \quad (17)$$

where mg is the weight of a copper ion and η is the viscosity of the solution. This gives a room temperature value $\gamma_c = 0.02$ for twice the radius of the thinnest wire used ($R = 25\ \mu$m) if no polymer is added. Because $\gamma_c \propto R^3$ this would leave less than one decade of radius before convection became important.

The addition of 1.4 wt% polyethylene oxide raised the overall viscosity by a factor of ~1,000 without significantly affecting the diffusion of small inorganic ions. The range of diffusion-limited behaviour is thus extended in radius by an order of magnitude and the damping of mechanical disturbances also improved.

Conclusions

The two principal conclusions are that the electrolytic deposit of copper in experimental, diffusion-limited conditions is fractal, and that the Hausdorff dimensionality agrees closely with the simple computer model of diffusion-limited aggregation[2,5]. This is the first experimental verification of the conjecture by Witten and Sander that such fractal behaviour should occur at long length scales irrespective of local details. The experimental deposits were highly ordered up to order 40 nm, but the Hausdorff dimensionality was measured over the range 25–400 μm. Extrapolating the experimental data and comparing with electron micrograph observations suggests that in fact the fractal behaviour extends over the full range 40–400,000 nm.

There are several possible sources of systematic error in the present experiment. Convection, agitation, contact resistance and resistance and rigidity of the deposit were all problems that we believe were brought under control. Other possible problems include non-inertness of the polymer, and the presence of any surface-active impurities.

In future work there are four aspects of the experiment which could be improved or varied: the type of measurement made; the electrolyte solution; the geometry; and the physical regime.

That the surface area of the deposit is proportional to its mass could be verified by measuring the capacitance. If the screening length in the solution is short and resistances can be rendered small or corrected for, the capacitance should be proportional to surface area.

The complications associated with the two possible oxidation states of copper might be avoided by, for example, the use of silver electrolytes. Convection might be suppressed without resort to polymer by the use of aqueous Ni^{2+} in the presence of excess CsCl: with these species there is the possibility of arranging that there be no net density change of the solution near the cathode. Other viscosity enhancers might also be tried.

It might be possible to make measurements in a two-dimensional geometry by, for example, using filter paper soaked in the electrolyte. In three-dimensional experiments, the use of microelectrodes would allow measurements to be made down to smaller length scales, but there could be mechanical problems with such an arrangement. A novel and more stable geometry would be to grow from a 'pinprick' (perhaps made by a spark) in the insulation on a plate: it would be useful to perform the corresponding computer simulation for growth from a point seed into the half space above a plane.

In our experiments the deposits grew slowly compared with the diffusion of ions. It should be possible by using higher electrolyte concentrations to explore the region where the deposits grow more quickly and behaviour is controlled by the diffusion length[1,10]. An alternative regime leading to fractal deposition might be obtained by stirring the solution until the reaction becomes limited not by diffusion but by ohmic resistance. It would then be the equations of electrostatics which drive the MS instability at long length scales.

R.C.B. and R.M.B. acknowledge Research Fellowships from Gonville and Caius College and from Trinity College, Cambridge respectively. We also thank Union Carbide Ltd for supplying their WSR 301 Polyox, the polymer used in our experiments. Finally we thank Drs J. Agar, M. Archer, J. Deutsch and P. Luckham for valuable advice and discussions.

Received 21 October 1983; accepted 13 March 1984.

1. Langer, J. *Rev. Mod. Phys.* **52**, 1–28 (1980).
2. Witten, T. A. & Sander, L. M. *Phys. Rev. Lett.* **47**, 1400–1403 (1981).
3. Mandelbrot, B. B. *The Fractal Geometry of Nature* (Freeman, San Francisco, 1977).
4. Hausdorff, F. *Math. Ann* **79**, 157, 179 (1919).
5. Meakin, P. *Phys. Rev.* **A27**, 604–607, 1495–1507 (1983).
6. Forrest, S. R. & Witten, T. A. Jr *J. Phys.* **A12**, L109–L117 (1979).
7. von Smoluchovski, M. *Phys. Z.* **17**, 557–571, 585–599 (1916); *Z. Phys. Chem.* **92**, 129–143 (1917).
8. *Handbook of Chemistry and Physics* 60th edn (CRC Press, 1980).
9. *Smithsonian Physical Tables* 9th edn 398–399 (1954).
10. Nauenberg, M., Richter, R. & Sander, L. M. Preprint NSF ITP 83-60 (1984).

Reprinted with permission from the *J. of Phys. A: Math. Gen.* **10**, H.E. Stanley, "Cluster Shapes at the Percolation Threshold: An Effective Cluster Dimensionality and its Connection with Critical-Point Exponents," pp. L211- L220, ©1977 IOP Publishing Ltd., Bristol, UK.

LETTER TO THE EDITOR

Cluster shapes at the percolation threshold: an effective cluster dimensionality and its connection with critical-point exponents†

H Eugene Stanley

Department of Physics, Boston University, Boston, Massachusetts 02215, USA

Received 21 June 1977, in final form 26 September 1977

Abstract. An effective dimensionality d_p is introduced for the purpose of providing a quantitative characterisation of the degree of ramification of the clusters that occur at the percolation threshold. It is found that d_p is directly related to percolation critical exponents, and that $1 \leq d_p \leq d$, which in turn places bounds on certain scaling powers and critical exponents. The exponents—when renormalised according to Suzuki's 'extended universality' prescription—have an appealingly simple form in terms of d_p; in particular, the renormalised mean cluster size exponent is d_p, while both the order parameter and 'decay of correlation' exponents are given by the co-dimension $d - d_p$.

1. Introduction

The possibility that cluster shapes near the critical point are characterised by a high degree of ramification or 'stringiness' has been discussed frequently in recent work (see, e.g., Domb *et al* 1975, Stauffer 1975, Quinn *et al* 1976, Binder 1976, Temperley 1976, Leath 1976, Domb 1976). However the precise significance of this concept is yet to be clarified, and attempts to place the qualitative feature of ramification on a fully quantitative footing have been fraught with problems. It is plausible that some features of cluster shape may be directly related to aspects of observed critical phenomena; therefore it is desirable to identify certain properties of cluster shape that have precise geometric meaning and are related unambiguously to the quantitative parameters currently used to characterise the critical point. In this Letter we propose such a property, an 'effective cluster dimensionality' d_p‡.

We focus our discussion here on cluster shapes near the connectivity threshold, the 'critical point' of the percolation problem; the generalisation to other systems is under investigation. To a first approximation, the percolation problem is a useful description of the condensation of f-functional monomers at the gelation threshold (Flory 1953 and references therein). In particular, the sudden condensation of a single gel molecule—distinguished from the molecules of the sol phase by being essentially infinite in spatial extent—is analogous to the appearance of a single infinite cluster at p_c in the

† Work supported by the NSF and the AFOSR.

‡ So far as we can tell, d_p is equivalent to the Hausdorff–Besicovitch dimensionality (Hausdorff 1919, Besicovitch 1935), a mathematical concept further developed and extensively applied to a wide range of systems by Mandelbrot (1975); Mandelbrot calls objects with non-integer Hausdorff–Besicovitch dimensionality 'fractals', and the Hausdorff–Besicovitch dimensionality itself he terms the 'fractal dimensionality'.

percolation problem. Thus to the extent that this analogy is valid, we may interpret d_p as an 'effective polymer dimensionality' at the gelation threshold.

We will find that d_p is significantly smaller than d, the system dimensionality, which is consistent with the qualitative notion that clusters near the critical point are highly ramified. In particular, we find $d_p = 2$ for the Flory theory of gelation, implying that in this closed form theory the molecules are effectively two dimensional even though they may 'fill' a three-dimensional space.

2. Definition of d_p

If the incipient infinite cluster at p_c had the full connectivity of a d-dimensional lattice, then we would expect that for large finite clusters very near p_c, $\chi_p \sim \xi_p^d$; here χ_p denotes the mean number of occupied sites belonging to a cluster, and ξ_p denotes the mean linear dimension of a cluster†. For highly ramified clusters, on the other hand, we expect $\chi_p \ll \xi_p^d$ (cf figure 1). It is of interest to enquire whether there exists a number d_p such that

$$\chi_p \sim (\xi_p)^{d_p}. \tag{1a}$$

Indeed, such a number would seem to exist, since from the definitions of the critical exponents γ_p and ν_p characterising the divergence at p_c of χ_p and ξ_p respectively, we have

$$d_p = \gamma_p / \nu_p. \tag{1b}$$

Note that the 'effective cluster dimensionality' d_p defined in (1*a*) should depend only on d since percolation exponents depend only on d.

3. Dependence of d_p on d

For $d = 1$ and $d = \infty$, we know the critical exponents exactly and hence we can calculate d_p exactly‡. We find

$$d_p(1) = 1, \tag{2a}$$

which is not surprising, and

$$d_p(\infty) = 2, \tag{2b}$$

† We use the definitions of percolation functions and corresponding critical exponents exactly as in Essam (1972) with one exception: we use the notation χ_p instead of $S(p)$ in order to stress the parallel with thermal phase transitions—for example, for $d = 1$, $\chi_p = (1+p)/(1-p)$ while $\chi_T = (1+y)/(1-y)$, where χ_T is the isothermal susceptibility and $y \equiv \tanh(J/kT)$ (Reynolds *et al* 1977, Klein *et al* 1977, Stauffer and Jayaprakash 1977, Stanley *et al* 1977).

‡ For $d = 1$, $p_c = 1$, $\chi_p \sim (1+p)/(1-p)$, and $\xi_p^2 = 2p/(1-p)^2$; hence $\gamma_p = \nu_p = 1$. The lattice dimensionality of a Cayley tree is not well defined (it cannot be embedded in any finite-dimensional space), but it is believed that percolation exponents for the interior of a Cayley tree (Flory 1953, Fisher and Essam 1961) are those of an infinite-dimensional system. For a Cayley tree, there are no closed cycles, and the Flory theory of the gelation threshold becomes exact for this lattice; consequently the dimensionality of the incipient gel molecule is two, regardless of the system dimensionality (cf figure 2).

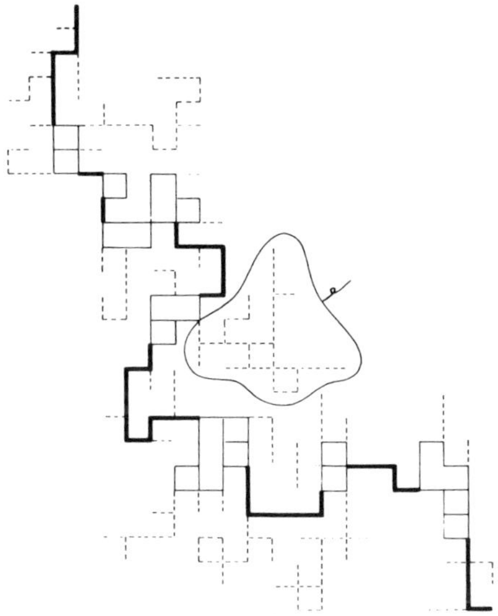

Figure 1. Computer-generated 'incipient infinite cluster' for the bond percolation problem on a 25×20 square lattice with $p = p_c = 0{\cdot}5$. Note the qualitative feature of 'ramification' or 'stringiness' of this cluster, corresponding to the fact that the effective cluster dimensionality d_p is less than d (for $d = 2$, $d_p \simeq 1{\cdot}8$). If we think of this cluster as representing a dilute magnetic system in which only a fraction p of the exchange integrals are non-zero, then it is clear that magnetic correlations will spread from one end of the lattice to another along a path that is dominated by the 'backbone bonds' (shown as full lines) rather than by the 'dangling ends' (shown as broken lines). The backbone bonds that are singly-connected are shown as bold lines while those that are multiply connected are shown as light lines. By construction, the bold lines form self-avoiding walks, and it has been conjectured that 'globally' the backbone is a SAW interpreted by 'blobs' of multiply-connected bonds. In support of this conjecture, d_p is identical for both the percolation problem and the SAW problem.

which is surprising since it seems to imply that in the limit of infinite lattice dimensionality the effective clüster dimensionality is only two!

What do we expect for $1 < d < \infty$? For $d = 1$, our intuitive understanding of equation (2*a*) is that the connectivity necessary for percolation is achieved only for $p = 1$, and the percolation cluster therefore has $d_p = d$. For $d > 1$, we expect that the percolation cluster may have $d_p < d$, since percolation is achieved at $p_c < 1$. We also expect $d_p > 1$, since the cluster shape at p_c will be considerably more 'branched' than a simple linear chain. Thus for $d > 1$, we expect that $1 < d_p < d$.

Although no exact results for percolation exponents are known for $1 < d < \infty$, considerable information on cluster geometry has been obtained using the methods of low-density and high-density series expansions, direct computer simulations, position-space renormalisation group, and renormalisation group expansions in the parameter $(6-d)$. The relevant results of this work are summarised in table 1. From the limited information on γ_p and ν_p it is impossible to obtain accurate estimates of d_p, especially for $d = 4$, 5. However, if we make the plausible assumption that weak scaling holds (Essam and Gwilym 1971), then we can use the additional expression

$$d_p = (2 - \alpha_p - 2\beta_p)/\nu_p. \tag{3}$$

Table 1. Summary of existing calculations that can be used to obtain numerical values for $d_p(d)$. Each line of the table represents a different estimate, obtained using the method shown in the first column and the critical exponent values shown in subsequent columns. Here PSRG denotes 'position space renormalisation group'.

Method and equation used		γ_p	ν_p	α_p	β_p	δ_p	d_p
$d=2$							
series, equation	(1*b*)	2·38 ± 0·02 [a]	1·34 ± 0·02 [a]	—	—	—	1·78 ± 0·04
	(1*b*)	2·43 ± 0·03 [b]	1·34 ± 0·02 [a]	—	—	—	1·81 ± 0·05
	(1*b*)	2·43 ± 0·03 [b]	$1{\cdot}32\{^{+0{\cdot}02}_{-0{\cdot}07}$ [c]	—	—	—	$1{\cdot}84\{^{+0{\cdot}13}_{-0{\cdot}05}$
	(1*b*)	2·38 ± 0·02 [a]	$1{\cdot}32\{^{+0{\cdot}02}_{-0{\cdot}07}$ [c]	—	—	—	$1{\cdot}80\{^{+0{\cdot}12}_{-0{\cdot}04}$
	(3)	—	1·34 ± 0·02 [a]	−0·668 ± 0·004 [d]	0·138 ± 0·007 [e]	—	1·785 ± 0·04
	(3)	—	$1{\cdot}32\{^{+0{\cdot}02}_{-0{\cdot}07}$ [c]	−0·668 ± 0·004 [d]	0·138 ± 0·007 [e]	—	$1{\cdot}81\{^{+0{\cdot}12}_{-0{\cdot}04}$
	(4*a*)	2·38 ± 0·02 [a]	—	−0·668 ± 0·004 [d]	—	—	1·784 ± 0·02
	(4*a*)	2·43 ± 0·03 [b]	—	−0·668 ± 0·004 [d]	—	—	1·82 ± 0·02
	(4*b*)	2·38 ± 0·02 [a]	—	—	0·138 ± 0·007 [e]	—	1·792 ± 0·04
	(4*b*)	2·43 ± 0·03 [b]	—	—	0·138 ± 0·007 [e]	—	1·796 ± 0·05
	(4*c*)	—	—	−0·668 ± 0·004 [d]	0·138 ± 0·007 [e]	—	1·793 ± 0·016
	(4*d*)	—	—	—	—	18·0 ± 0·75 [f]	$1{\cdot}79\{^{+0{\cdot}16}_{-0{\cdot}14}$
computer simulations		—	—	—	—	—	1·78 [g]
PSRG: square; bond		—	—	—	—	18·249 [h]	1·792
PSRG: square; bond		—	—	—	—	—	1·9 [i]
PSRG: triangular; site		—	—	—	—	—	1·68 [i]
RG (Migdal–Kadanoff): square; bond		—	—	—	—	—	1·78 [j]
ϵ expansion	(5)	—	—	—	—	—	2·55
$d=3$							
series, equation	(1*b*)	1·66 ± 0·07 [k]	$0{\cdot}83 + 15\Delta p_c$ ± 0·01 [a]	—	—	—	2·00 ± 0·11
	(1*b*)	1·66 ± 0·07 [k]	$0{\cdot}825 \pm 50\Delta p_c$ ± 0·02 [c]	—	—	—	2·01 ± 0·22
	(1*b*)	1·70 ± 0·11 [a]	$0{\cdot}825 + 50\Delta p_c$ ± 0·02 [c]	—	—	—	2·06 ± 0·27
	(4*b*)	1·66 ± 0·07 [k]	—	—	0·42 ± 0·06 [l]	—	$1{\cdot}992\{^{+0{\cdot}25}_{-0{\cdot}23}$
	(4*b*)	1·70 ± 0·11 [a]	—	—	0·42 ± 0·06 [l]	—	$2{\cdot}008\{^{+0{\cdot}34}_{-0{\cdot}30}$
	(4*d*)	—	—	—	—	5·0 ± 0·8 [m]	$2{\cdot}00\{^{+0{\cdot}76}_{-0{\cdot}59}$
computer	(4*b*)	1·8 ± 0·005 [n]	—	—	0·39 ± 0·02 [n]	—	$2{\cdot}09\{^{+0{\cdot}14}_{-0{\cdot}12}$
ϵ expansion	(5)	—	—	—	—	—	2·34
$d=4$							
series	(4*b*)	1·41 ± 0·25 [o]	—	—	0·52 ± 0·03 [n]	—	$2{\cdot}30\{^{+0{\cdot}80}_{-0{\cdot}62}$
computer	(4*b*)	1·6 ± 0·1 [n]	—	—	0·52 ± 0·03 [n]	—	$2{\cdot}43\{^{+0{\cdot}32}_{-0{\cdot}28}$
ϵ expansion	(5)	—	—	—	—	—	2·18

Table 1.—*continued*

Method and equation used		γ_p	ν_p	α_p	β_p	δ_p	d_p
$d=5$							
series	(4*b*)	$1\cdot25\pm0\cdot15$[o]	—	—	$0\cdot66\pm0\cdot05$[n]	—	$2\cdot43\{^{+0\cdot58}_{-0\cdot48}$
computer	(4*b*)	$1\cdot3\pm0\cdot1$[n]	—	—	$0\cdot66\pm0\cdot05$[n]	—	$2\cdot48\{^{+0\cdot41}_{-0\cdot36}$
ϵ expansion	(5)	—	—	—	—	—	2·07
$d=6$							
series	(4*b*)	$1\cdot06\pm0\cdot20$[o]	—	—	$0\cdot97\pm0\cdot05$[n]	—	$2\cdot12\{^{+0\cdot68}_{-0\cdot56}$
computer	(4*b*)	$1\cdot00\pm0\cdot05$[n]	—	—	$0\cdot97\pm0\cdot05$[n]	—	$2\cdot04\{^{+0\cdot22}_{-0\cdot20}$
ϵ expansion	(5)	—	—	—	—	—	2·0

[a] Dunn *et al* (1975): *b*
[b] Sykes *et al* (1976b): *b,s*
[c] Cox and Essam (1976): *s*
[d] Domb and Pearce (1976): *b*
[e] Sykes *et al* (1976c): *b,s*
[f] Gaunt and Sykes (1976): *s,b*
[g] Mandelbrot (1977): *s*
[h] Dasgupta (1976): *b*
[i] Reynolds *et al* (1977): *b,s*
[j] Riedel *et al* (1977): *b*
[k] Sykes *et al* (1976d): *s*
[l] Sykes *et al* (1976*a*): *b,s*
[m] Gaunt (1977): *s*
[n] Kirkpatrick (1976): *s*
[o] Gaunt *et al* (1976): *s,b*

Further, if we believe that $d\nu_p \geq 2-\alpha_p$, then we obtain four additional expressions for d_p:

$$d_p \leq \begin{cases} d\gamma_p/(2-\alpha_p) & (4a) \\ d\gamma_p/(2\beta_p+\gamma_p) & (4b) \\ d(2-\alpha_p-2\beta_p)/(2-\alpha_p) & (4c) \\ d(\delta_p-1)/(\delta_p+1) & (4d) \end{cases}$$

where (4*d*) follows from the three-exponent equality $\gamma_p=\beta_p(\delta_p-1)$. All inequalities become equalities if strong scaling ('hyperscaling') holds, as is widely believed to occur at least for $d=2$, 6. The numerical data of table 1 are consistent with the result that $d_p<d$; in particular, it appears that $d_p(2)\simeq 1\cdot8$ and $d_p(3)\simeq 2$.

The situation for $d\simeq 6$ is particularly interesting. Firstly, Toulouse (1974) has conjectured that the Bethe lattice results hold for $6\leq d<\infty$, and his conjecture has received support from numerical work of Kirkpatrick (1976) and Gaunt *et al* (1976). Thus it is plausible that $d_p(d)=2$ for $d>6$. Harris *et al* (1975) have proposed that for $d\lesssim 6$,

$$d_p(\epsilon)=2+\epsilon/21+(206/3^3 7^3)\epsilon^2+O(\epsilon^3) \qquad (5)$$

where $\epsilon\equiv 6-d$; the order ϵ^2 term was calculated by Priest and Lubensky (1976) and Amit (1976). Since as d increases, the branching possibilities should not decrease, one would intuitively expect that d_p is a non-decreasing function of d‡. If this is the case, then $d_p(\infty)=2$ is the maximum value of d_p. Therefore equation (5) is surprising: although it predicts $d_p<d$ for all reasonable ϵ, it also predicts that the cluster dimensionality *increases* as the system dimensionality decreases (cf figure 2).

In summary, then, all available data are consistent with the conjecture that $d_p<d$.

‡ This intuition may be false, since $p_c(d)$ decreases with increasing d (P J Reynolds, private communication).

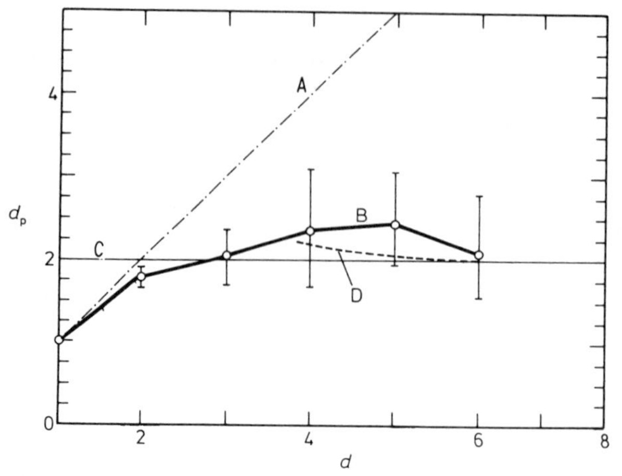

Figure 2. Shown are four possible dependences of d_p on d. A, if the incipient infinite cluster at p_c (or, equivalently, the incipient gel molecule) had the full connectivity of a d-dimensional lattice, then we would expect that $d_p = d$. B, for $d > 1$, the open circles represent averages of the numerical estimates of d_p from table 1 (and the 'error bars' indicate the range of the estimates), while for $d = 1$ there exists an exact solution. Note that $d_p \leqslant d$, and the difference $d - d_p$ (the co-dimension) is the Suzuki-renormalised order parameter exponent $2\beta_p/\nu_p$ and the 'decay of correlation' exponent $d - 2 + \eta_p$. C, the Flory theory of the gelation threshold predicts that $d_p = 2$ *for all* d; this result is correct in the limit $d \to \infty$, and probably correct for $d \geqslant 6$ (Toulouse 1974), but it is geometrically impossible for $d < 2$. Note that for $d = 3$, the Flory theory is in good agreement with numerical predictions. D, the $(6 - d)$ expansion proposed by Harris *et al* (1975) for $d \lesssim 6$.

4. Connection with the scaling hypothesis

The scaling hypothesis for the percolation functions which are analogous to thermodynamic functions was first formulated by Essam and Gwilym (1971). Here it is convenient to consider the scaling hypothesis directly for the pair connectedness $C_2(r, \Delta p, h)$, where h is the analogue of the magnetic field for the percolation problem, and $\Delta p \equiv p - p_c$. One assumes there exist three numbers b_q, b_p, and b_h such that for all positive λ,

$$S(\lambda^{b_q} q, \lambda^{b_p} \Delta p, \lambda^{b_h} h) = \lambda^{1 - d b_q} S(q, \Delta p, h) \tag{6}$$

where $S(q)$ is the spatial Fourier transform of $C_2(r)$. From (6) it follows that all critical exponents for functions derived from the pair connectedness can be expressed in terms of the three scaling powers b_q, b_p and b_h. In particular, one finds that $\nu_p = b_q/b_p$, $-\gamma_p = (1 - d b_q)/b_p$, and $2 - \eta_p = d - b_q^{-1}$ in complete analogy to the thermal problem (Hankey and Stanley 1972). Hence

$$d_p = 2 - \eta_p \tag{7}$$

and the expression $2 - \eta_p$ that occurs so frequently in relations among critical exponents is given a concrete interpretation in terms of a specific 'geometric' property of the percolation threshold. While $2 - \eta_p$ is the effective dimensionality, the length scaling power b_q^{-1} is the effective *co-dimensionality*,

$$b_q^{-1} = d - d_p. \tag{8}$$

The above relations follow directly from the 'weak scaling' assumption (6). If one further makes the 'strong scaling' assumption, then it follows that there are not three but only two independent scaling powers, since $b_h = db_q - \frac{1}{2}$ (Hankey and Stanley 1972). Hence we can also relate b_h to d_p: $2b_h = (d+d_p)/(d-d_p)$. Equivalently, $2y_h = d + d_p$, where $y_h = da_h$ is the usual magnetic field scaling power that occurs in thermodynamic scaling (a_h is defined precisely in Hankey and Stanley 1972). Therefore the inequality $1 \leq d_p \leq d$ leads to bounds on the scaling powers:

$$0 \leq b_q^{-1} \leq d-1, \qquad \frac{d+1}{d-1} \leq 2b_h \leq \infty, \qquad d+1 \leq 2y_h \leq 2d, \tag{9a}$$

which in turn lead to inequalities on the numerical values of the critical exponents that depend only on the length (or 'field') scaling powers:

$$2-d \leq \eta_p \leq 1 \quad \text{and} \quad \frac{d+1}{d-1} \leq \delta_p \leq \infty. \tag{9b}$$

Note that for the case $d=1$, $1 \leq d_p \leq d$ becomes a strict equality, and we have, from (9), $b_q^{-1} = 0$, $b_h = \infty$, $y_h = 1$, $\eta_p = 1$, and $\delta_p = \infty$.

5. Connection with the extended universality hypothesis

Suzuki (1974) has pointed out that the apparent exceptions to the universality hypothesis are in fact not exceptions if one defines all critical exponents not through the variables q, H, and $\Delta T \equiv T - T_c$ but rather through the variables q, H, and $\kappa \sim (\Delta T)^{\nu}$. If we carry this procedure over to the percolation problem, equation (6) becomes

$$S(\lambda q, \lambda\kappa, \lambda^{b_h/b_q} h) = \lambda^{-d_p} S(q, \kappa, h) \tag{10}$$

and the number of independent scaling powers is reduced from three to two†. It is interesting to note that if we define a second 'fractal dimensionality' $d_\sigma = (2-\alpha_p)/\nu_p$, then both scaling powers and hence all critical exponents are readily expressible as simple functions of the two fractal dimensionalities d_σ and d_p. For example, the 'Suzuki-renormalised' order parameter (or 'weight fraction') exponent $\hat{\beta} \equiv 2\beta_p/\nu_p$ becomes $d_\sigma - d_p$, the mean size exponent $\hat{\gamma} = \gamma_p/\nu_p$ becomes d_p, the free energy exponent $\phi \equiv (2-\alpha_p)/\nu_p$ becomes d_σ, and the 'critical isotherm' exponents $\hat{\delta} \equiv \delta_p$ and $\hat{\eta} \equiv (d-2+\eta_p)$ become $(d_\sigma + d_p)/(d_\sigma - d_p)$ and $d - d_p$ respectively.

The numerical value of d_σ remains unspecified under the weak scaling assumption (one might conjecture $d_\sigma \leq d$ by analogy with the Josephson inequality, and $d_\sigma \geq d_p$ if $\beta_p \geq 0$). However, if we make the additional assumption necessary for 'strong scaling', then $b_h = db_q - \frac{1}{2}$; hence $d_\sigma = d$ and there is only a single independent parameter, d_p, in terms of which all renormalised critical exponents find remarkably simple expression: $\hat{\gamma} = d_p$, $\hat{\beta} = \hat{\eta} = d - d_p$ (the co-dimension), $\hat{\delta} = (d+d_p)/(d-d_p)$, and $\phi = d$.

Thus any two systems with identical values of d_p will have identical values of all 'renormalised' critical exponents provided strong scaling holds. For example, Suzuki observed that the eight-vertex model—which describes two independent Ising models on square lattices coupled by a four-spin interaction—has the same values of all renormalised critical exponents, independent of the coupling strength. This observation can be interpreted as the intuitively plausible statement that for this system the

† (10) follows on substituting $\kappa = (\Delta p)^{b_q/b_p}$ into (6), multiplying all scaling powers by b_q^{-1}, and using (8).

effective cluster dimensionality (or degree of cluster 'ramification') is independent of the coupling strength. Suzuki has also pointed out that the renormalised critical exponents are the same for the Ising model with triplet interactions as for the Ising model with pair interactions; this result can be interpreted to mean that both systems have the same effective cluster dimensionality. Finally, it should be pointed out that if two systems have the same value of d_p, there may be other common features. For example, Stanley *et al* (1976) have pointed out that d_p appears to be the same for the percolation and self-avoiding walk (SAW) problems, and this result has been interpreted to mean that correlations spread throughout a dilute magnet at the percolation threshold along paths that are, globally, SAW's (cf figure 1).

6. Discussion

Thus far our attention has focused on the particular definition of d_p given in equation (1). It is certainly legitimate to ask what would happen if we were to adopt differing definitions of d_p. Logically, we can alter equation (1) by changing either the length or the size functions therein. There is no *a priori* basis for choosing a 'best' definition of the effective cluster dimensionality, and definitions different than (1) may prove more suitable for certain applications. One alternative to (1)—which is equivalent to (1) when hyperscaling holds but gives larger numerical values for the effective cluster dimensionality—is obtained by focusing on the very large clusters. The typical length of the very large clusters diverges with the same power of (p_c-p), but the typical size s_L of the very large clusters may diverge faster than the mean size $\chi_p \equiv \langle s \rangle$ of all the finite clusters, thus leading to a different value of d_p. Information about s_L can be obtained by the following argument (Stauffer 1976): assume that the mean number of s-site clusters varies with s and p according to

$$n_s(p) \sim s^{-\tau} f((p_c-p)s^{1/\beta\delta}) \tag{11}$$

for very large s and very small $(p_c-\mathrm{p})/p_c$; then in some sense a typical size s_L of the very large clusters described by (11) may be expected to diverge as $s_L \sim (p_c-p)^{-\beta\delta}$. Note that for $d>1$, s_L grows much more rapidly as $p \to p_c$ than $\langle s \rangle$, since $\beta\delta = \beta + \gamma > \gamma$. Defining $d_p^\dagger$ through $s_L \sim \xi_p^{d_p^\dagger}$, we have†

$$d_p^\dagger = \beta_p \delta_p / \nu_p = d_p[\delta_p/(\delta_p - 1)] \leqslant (d+d_p)/2. \tag{12}$$

If strong scaling holds, the inequality becomes an equality and $d_p^\dagger$ is fully equivalent to d_p, while if strong scaling does not hold, $d_p^\dagger$ and d_p are related by a critical point exponent. That the mean of a sub-group differs from the mean of the entire group is not surprising; what is perhaps intriguing is that as $p \to p_c$ the very large clusters increase in size sufficiently faster than the mean clusters that their effective dimensionality $d_p^\dagger$ is larger (for $d>1$) than the effective dimensionality d_p of the system as a whole.

It is straightforward to derive numerical values for $d_p^\dagger$ from the information given above and in table 1. In particular, we have $d_p^\dagger(d) \geqslant d_p(d)$, with the equality holding

† Very recently Harrison, Bishop and Quinn (private communication, and to be published) have obtained expression (12) for $d_p^\dagger$ using an elegant argument that does not explicitly involve the cluster-size scaling hypothesis. Moreover, they have confirmed (12) numerically for $d=2$, 3 by direct computer simulation of large clusters near p_c. Earlier $d=2$ data of Leath (1976) on smaller samples are also consistent with (12) (Leath interprets his data as being consistent with a fractal dimensionality of two).

for $d = 1$, and (from (5))

$$d_{\mathrm{p}}^{\dagger}(\epsilon) = 4 - \frac{10}{21}\epsilon + \frac{103}{3^3 7^3}\epsilon^2 + \dots \tag{13}$$

which results in the *monotonic* sequence $d_{\mathrm{p}}^{\dagger}(\epsilon) = 4$, 3·53, 3·09, 2·67 for $d = 6, 5, 4, 3$ respectively. If strong scaling holds, then $2b_h = d_{\mathrm{p}}^{\dagger}/(d - d_{\mathrm{p}}^{\dagger})$ and $y_h = d_{\mathrm{p}}^{\dagger}$, while if strong scaling does not hold, no scaling power is given in terms of $d_{\mathrm{p}}^{\dagger}$ alone. The 'Suzuki-renormalised' critical exponents, when expressed in terms of $d_{\mathrm{p}}^{\dagger}$, are $\hat{\beta} = \hat{\eta} = 2(d - d_{\mathrm{p}}^{\dagger})$, $\hat{\gamma} = 2d_{\mathrm{p}}^{\dagger} - d$, and $\hat{\delta} = d_{\mathrm{p}}^{\dagger}/(d - d_{\mathrm{p}}^{\dagger})$.

In conclusion, then, we have proposed an effective cluster dimensionality d_{p} that describes the shape of the clusters (molecules) present at the percolation (gelation) threshold. We showed that d_{p} is directly related to critical exponents and we then used existing information about these exponents to evaluate d_{p}; d_{p} depends only on d, and $1 \leq d_{\mathrm{p}} \leq d$ for all systems considered (with the equality holding only for $d = 1$), which is consistent with the intuitive notion that the incipient gel molecule is highly ramified. Finally, we made connection with the scaling and extended universality hypotheses. In particular, we found that the inequality $1 \leq d_{\mathrm{p}} \leq d$ places bounds on the scaling powers and certain critical exponents, and that Suzuki's renormalised critical exponents all assume remarkably simple expressions when written in terms of d_{p}; in particular, the mean size exponent is simply d_{p} and the order parameter and 'decay of correlation' exponents are simply the co-dimension $d - d_{\mathrm{p}}$.

Acknowledgments

We wish to thank P J Reynolds, S Redner, and W Klein for many useful discussions throughout the course of this work that helped clarify many of the points involved. The remarks on $d_{\mathrm{p}}^{\dagger}$ were formulated after extensive discussion also with G Bishop, R J Harrison, and D Stauffer. We also wish to thank B Mandelbrot, J F Nicoll, G F Tuthill, R J Birgeneau, and M E Fisher for helpful comments on this work.

References

Amit D J 1976 *J. Phys. A: Math. Gen.* **9** 1441–59
Besicovitch A S 1935 *Math. Annln* **110** 321–32
Binder K 1976 *Ann. Phys., NY* **98** 390–417
Cox M A A and Essam J W 1976 *J. Phys C: Solid St. Phys.* **9** 3985–91
Dasgupta C 1976 *Phys. Rev.* B **14** 1221–4
Domb C 1976 *J. Phys. A: Math. Gen.* **9** 283–99
Domb C and Pearce C J 1976 *J. Phys. A: Math. Gen.* **9** L137–40
Domb C, Schneider T and Stoll E 1975 *J. Phys. A: Math. Gen.* **8** L90–4
Dunn A G, Essam J W and Ritchie D S 1975 *J. Phys. C: Solid St. Phys.* **8** 4219–35
Essam J W 1972 *Phase Transitions and Critical Phenomena* vol. 2, eds C Domb and M S Green (London: Academic) chap. 6
Essam J W and Gwilym K M 1971 *J. Phys. C: Solid St. Phys.* **4** L228–31
Fisher M E and Essam J W 1961 *J. Math. Phys.* **2** 609–19
Flory P J 1953 *Principles of Polymer Chemistry* (Ithaca: Cornell University Press)
Gaunt D S 1977 *J. Phys. A: Math. Gen.* **10** 807–12
Gaunt D S and Sykes M F 1976 *J. Phys. A: Math. Gen.* **9** 1109–16
Gaunt D S, Sykes M F and Ruskin H 1976 *J. Phys. A: Math. Gen.* **9** 1899–911

Hankey A and Stanley H E 1972 *Phys. Rev.* B **6** 3515–42
Harris A B, Lubensky T C, Holcomb W K and Dasgupta C 1975 *Phys. Rev. Lett.* **35** 327–30
Hausdorff F 1919 *Math. Annln* **79** 157–79
Kirkpatrick S 1976 *Phys. Rev. Lett.* **36** 69–72
Klein W, Stanley H E, Redner, S and Reynolds P J 1977 *Boston University Preprint*
Leath P L 1976 *Phys. Rev.* B **14** 5046–55
Mandelbrot B 1975 *Les Objects Fractals* (Paris, Montreal: Flammarion)
—— 1977 to be published
Priest R G and Lubensky T C 1976 *Phys. Rev.* B **13** 4159–71
Quinn G D, Bishop G H and Harrison R J 1976 *J. Phys. A: Math. Gen.* **9** L9–14
Reynolds P J, Klein W and Stanley H E 1977 *J. Phys. C: Solid St. Phys.* **10** L167–72
Reynolds P J, Stanley H E and Klein W 1977 *J. Phys. A: Math. Gen.* **10** L203–9
Riedel E K, Jayaprakash C and Wortis M 1977 *Bull. Am. Phys. Soc.* **22** 301
Stanley H E, Birgeneau R J, Reynolds P J and Nicoll J F 1976 *J. Phys. C: Solid St. Phys.* **9** L553–60
Stanley H E, Klein W and Reynolds P J 1977 *Boston University Preprint*
Stauffer D 1975 *J. Phys. C: Solid St. Phys.* **8** L172–7
—— 1976 *Z. Phys.* B **25** 391–9
Stauffer D S and Jayaprakash C 1977 *Preprint*
Suzuki M 1974 *Prog. Theor. Phys.* **51** 1992–3
Sykes M F, Gaunt D S and Essam J W 1976a *J. Phys. A: Math. Gen.* **9** L43–6
Sykes M F, Gaunt D S and Glen M 1976b *J. Phys. A: Math. Gen.* **9** 97–103
—— 1976c *J. Phys. A: Math. Gen.* **9** 725–30
—— 1976d *J. Phys. A: Math. Gen.* **9** 1705–12
Temperley H N V 1976 *J. Phys. A: Math. Gen.* **9** L113–7
Toulouse G 1974 *Nuovo Cim.* B **23** 234–40

Reprinted with permission from *Physical Review Letters* **53**, H.D. Bale and P.W. Schmidt, "Small-Angle X-Ray-Scattering Investigation of Sibmicroscopic Porosity with Fractal Properties," pp. 596–599, ©1984 American Physical Society.

VOLUME 53, NUMBER 6 PHYSICAL REVIEW LETTERS 6 AUGUST 1984

Small-Angle X-Ray-Scattering Investigation of Submicroscopic Porosity with Fractal Properties

Harold D. Bale[(a)] and Paul W. Schmidt
Physics Department, University of Missouri, Columbia, Missouri 65211
(Received 7 June 1984)

A method is developed for analyzing the outer part of the small-angle x-ray or neutron scattering curve for porous scatterers in which the pore boundaries can be described by fractals. When the results are applied to the scattering data from a lignite coal, the fractal dimension of the boundary surface of the pores in this coal is found to be 2.56 ± 0.03.

PACS numbers: 61.10.–i, 61.12.Dw, 68.20.+t, 68.45.By

A recent small-angle x-ray-scattering study of the submicroscopic porosity in several lignite coals[1] showed that there was a broad region of the scattering curve in which the scattered intensity was proportional to a nonintegral negative power of

$$q = 4\pi\lambda^{-1}\sin(\theta/2), \quad (1)$$

where λ is the x-ray wavelength and θ is the scattering angle. This angular dependence was very unexpected. since according to the traditional theories of small-angle x-ray and neutron scattering,[2] which assume that the scatterers are bounded by a smooth surface, the limiting form of the scattered intensity should be proportional to q^{-4}. After remeasuring the scattering and finding that our first results were reproducible, we ascribed[1] this power-law scattering to independently scattering submicroscopic pores with a power-law dimension distribution.[3]

After we submitted the manuscript of Ref. 1, however, we learned about the application of fractals[4] in studies of porosity.[5,6] We therefore have reexamined our results for lignite coals and have found that our data can alternatively be considered to be the scattering from micron-sized pores with boundary surfaces which are fractals.

Below, we outline some equations which we have developed for this analysis of scattering data in terms of fractals and discuss the application of these equations to the lignites which we have studied.

In the model which we have developed, the coal is considered to be a three-dimensional porous solid with uniform electron density δ. (The assumption of uniform electron density is valid for *small*-angle x-ray and neutron scattering, since at scattering angles no greater than a few degrees, the scattering process cannot resolve the atomic structure.[7]) The scattered intensity for this model can be calculated by a modification of a procedure[8] developed for analysis of the small-angle scattering from materials containing pores with smooth boundary surfaces. For a porous sample in which solid material with uniform electron density occupies a fraction c of the sample volume V, so that the pores take up a fraction $1-c$ of the sample volume, the scattered intensity $I(q)$ is given by[9]

$$I(q) = 4\pi I_e\delta^2 c(1-c)V\int_0^\infty r^2 g(r)\frac{\sin qr}{qr}dr. \quad (2)$$

In (2), I_e is the intensity scattered by a single electron; and $g(r)$, the correlation function, is defined by the equation

$$g(r) = [Z(r) - c]/(1-c),$$

where $Z(r)$ is the probability, averaged over all points of the sample and over all sample orientations, that if a point lies in an occupied region (i.e., *not* in a pore), a second point a distance r from the first point also will be in an occupied part of the sample.

To approximate $Z(r)$ for small r, we will use a technique similar to that employed by Guinier, Fournet, Walker, and Yudowitch,[10] in their calculation of the small-r approximation for $g(r)$ for a scatterer with uniform electron density and a smooth boundary surface. From the definition of the correlation function $Z(r)$ given above, $Z(r) = 1$ at all occupied points except those located within a distance r of a pore boundary. According to the definition of the fractal or Hausdorff dimension D of the fractal boundary surface,[11] as r approaches zero, the number n of cubes with edge r needed to make a layer covering all points of the pore-boundary surfaces can be expressed

$$n = N_0 r^{-D}, \quad (3)$$

where N_0 is a constant characteristic of the fractal boundaries. The layer of the occupied regions in which $Z(r) < 1$ thus has the volume

$$V_b = nr^3 = N_0 r^{3-D}, \quad (4)$$

while the volume of the sample in which $Z(r) = 1$ is equal to $cV - V_b$.

VOLUME 53, NUMBER 6 PHYSICAL REVIEW LETTERS 6 AUGUST 1984

For $x \leq r$, let $p(r,x)$ be the probability, averaged over all orientations of r, that if a point in the occupied region is located at a distance x from a pore boundary surface, there is another occupied point a distance r from the first point. If in our approximate calculation of $Z(r)$ for small r we assume that the boundary surface is a plane, then[10]

$$p(r,x) = (r+x)/2r. \tag{5}$$

Since $g(r)$ is defined to be the probability, averaged over both the inner region and the boundary layer, that two occupied points will be separated by a distance r, we find from (3), (4), and (5) that as r approaches zero,

$$Z(r) \simeq \frac{cV - V_b}{cV} + \frac{V_b}{cV}\frac{1}{r}\int_0^r p(r,x)\,dx$$

$$= 1 - N_0 r^{3-D}/4cV.$$

Thus, from the relation between $g(r)$ and $Z(r)$, we have

$$g(r) = 1 - N_0 r^{3-D}/4c(1-c)V. \tag{6}$$

Since we expect[12] that both $g(r)$ and $g'(r)$ approach zero as r becomes infinite, by two integrations of Eq. (2) by parts, $I(q)$ can be written

$$I(q) = -4\pi\delta^2 I_e c(1-c)Vq^{-3}\int_0^\infty [rg''(r) + 2g'(r)]\sin qr\,dr. \tag{7}$$

Erdélyi's theorem for asymptotic expansion of Fourier integrals[13] can be used to calculate the limiting form of $I(q)$ for large q. Thus

$$I(q) = -\frac{4\pi\delta^2}{q^{6-D}} I_e c(1-c)V\Gamma(3-D)\sin[\pi(D-1)/2]\lim_{r\to 0}\left\{r^{D-2}\frac{d^2}{dr^2}[rg(r)]\right\}.$$

From (6),

$$I(q) \simeq \pi N_0 \delta^2 I_e \Gamma(5-d)\sin[\pi(D-1)/2]/q^{6-D}. \tag{8}$$

According to this approximation, $I(q)$ is proportional to $q^{-(6-D)}$.

For a smooth pore-boundary surface, $D=2$, and from (3), $n = N_0 a^{-2}$. Comparison of (8) with the usual approximation[2] for $I(q)$ in the outer part of the small-angle scattering curve for samples containing scatterers with a smooth boundary surface shows that for $D=2$, N_0 can be shown to be equal to the total surface area S separating the two phases in the sample when $D=2$.

For a sample with mass M, the quantity N_0/M thus can be considered to be analogous to the specific surface S/M for pores with smooth boundaries. As

$$I_e N_0 = (I_e A)(M/A)(N_0/M), \tag{9}$$

and the quantities $I_e A$ and M/A can be evaluated by scattering measurements,[14] N_0/M can be calculated from the scattering data.

As (8) neglects the short-range atomic structure in the sample, this equation does not apply when q is large enough to permit resolution of structures with atomic dimensions. (The same restriction is true for most equations employed in the analysis of small-angle x-ray or neutron scattering.)

At least a rough estimate of the average pore dimension R can be obtained from $q_{\min}$, the smallest value of q at which Eq. (8) describes the scattering, since when this power-law scattering is observed, one expects $q_{\min}R \gg 1$. If as a conservative estimate we assume that $q_{\min}R > 3.5$, the average dimension of the pores must be at least $3.5/q_{\min}$.

Figure 1 shows the scattering curve which we obtained[1] for Beulah lignite. Information about the experimental conditions and data corrections is given in Ref. 1. For a wide interval of q, the intensity is proportional to $q^{-(3.44\pm0.03)}$. (The uncertainty in the exponent was estimated from the reproducibility of the scattering curve.) While, as we have mentioned, this kind of curve can be explained as being the result of the scattering from spherical pores with a power-law dimension distribution,[3] the scattering data can also be described by Eq. (8), which implies that the lignite contains pores which are bounded by surfaces with a fractal dimension $6-(3.44\pm0.03) = 2.56\pm0.03$. As the dependence on $q^{-(3.44\pm0.03)}$ holds even at the smallest angles at which scattering data can be registered, the pore dimension R must be at least 1 micron. However, in the outer part of the curve in Fig. 1, the scattering approaches a relatively low but constant value. We have interpreted[1] this part of the scattering curve as the result of density fluctuations with dimensions of the same magnitude as those of atoms or small molecules. In the least-squares fit from which we constructed the curve in Fig. 1, we used a fitting function of the form

$$I(q) = B + Cq^{-\alpha} \tag{10}$$

VOLUME 53, NUMBER 6 PHYSICAL REVIEW LETTERS 6 AUGUST 1984

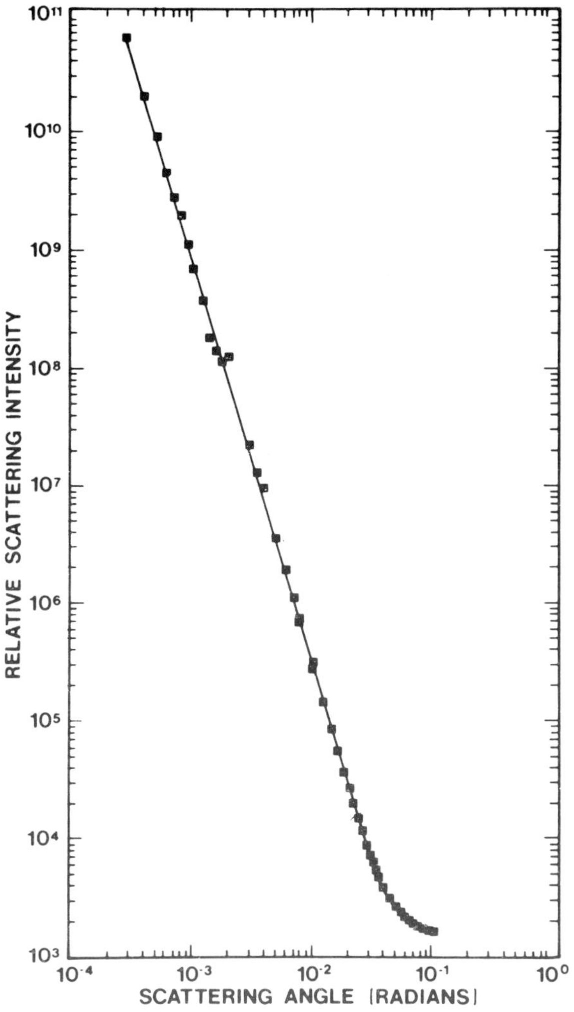

FIG. 1. The scattered intensity for Beulah lignite, measured with Bonse-Hart and Beeman collimation systems, as described in Ref. 1. The points show the corrected scattered intensities, and the curve was drawn by a least-squares fit of Eq. (10). [Reproduced by permission from "Chemistry of Low-Rank Coals," edited by H. D. Schobert, ACS Symposium Series (American Chemical Society, Washington, D.C., to be published).]

with the constants C, B, and α being determined in the fit. (The constant B was included in order to take account of the atomic-scale density fluctuations.) When $I_e A$ is evaluated from the x-ray scattering data by a technique similar to that employed in one of our previous small-angle scattering studies of coal porosity,[14] we find that for Beulah lignite $N_0/M = 20 \pm 2\ \mathrm{m}^{2.5}/\mathrm{g}$.

We have obtained similar scattering curves for all of the lignites which we have examined.[1] Although the apparatus did not allow the scattering to be measured over such a large interval of q as in the curve shown in Fig. 1, the scattered intensity for all the lignites and subbituminous coals which we studied was proportional to $q^{-\alpha}$ for a large interval of q, with $\alpha \simeq 3.5$. Kaiser and Gethner[5] found that the small-angle neutron scattering from Rawhide coal followed a power law for q between about 0.01 and 0.1 Å^{-1} and concluded[16] that the pores in this coal may have fractal properties. We therefore believe that the fractal interpretation applies to many lignites and other coals and that the fractal dimension of the pore boundary surfaces in these lignites is about 2.5.

Sinha, Freltoft, and Kjems[17] and Schaefer and Keefer[18] developed equations for analyzing their small-angle scattering data from fractal aggregates of silica particles. In their power-law equations for the scattered intensity, the exponent of q is not the same as in (8). These differences are to be expected, since fractal *objects* need not scatter in the same way as pores with fractal boundaries.

Pfeifer has pointed out[19] that systems of pores with fractal boundaries have many features similar to those of pore systems with a pore-dimension distribution function with the form of a power law. The *number* distribution corresponding to the *volume* distribution in Ref. 19 for the characteristic pore dimension a is proportional to $a^{-(D+1)}$, and the scattered intensity calculated[3] from this power-law distribution is proportional to $q^{-(6-D)}$, as given by Eq. (8). Systems of pores with a power-law dimension distribution thus scatter in essentially the same way for large q as pores with boundaries so rough that they can be described by fractal surfaces.

Avnir, Farin, and Pfeifer[5b] have shown that pores bounded by fractal surfaces can explain the adsorption data for several adsorbents. Their results suggest that such fractal-pore systems may be more common than was previously suspected. Perhaps much of published small-angle scattering data from porous samples can also be interpreted with Eq. (8). In the past, when we obtained a corrected scattering curve that in the outer portion was proportional to a negative power somewhat smaller in magnitude than 4, we assumed that the deviation was the result of some unidentified experimental effect. We now realize that there may be a more fundamental reason for these noninteger exponents.

We wish to express our appreciation to P. Pfeifer, M. Bretz, J. S. Gethner, S. K. Sinha, and L. M. Sander for their helpful suggestions and advice.

VOLUME 53, NUMBER 6 PHYSICAL REVIEW LETTERS 6 AUGUST 1984

This work was supported in part by the North Dakota Energy Research Center, and by the Donors of the Petroleum Research Fund administered by the American Chemical Society.

[(a)]Permanent address: Physics Department, University of North Dakota, Grand Forks, N.D. 57202.

[1]H. D. Bale, M. D. Carlson, M. Kalliat, C. Y. Kwak, and P. W. Schmidt, in "Chemistry of Low-Rank Coals," edited by H. D. Schobert, ACS Symposium Series (American Chemical Society, Washington, D.C., to be published).

[2]A. Guinier, G. Fournet, C. B. Walker, and K. L. Yudowitch, *Small-Angle Scattering of X-Rays* (Wiley, New York, 1955), p. 17, Eq. (26).

[3]P. W. Schmidt, J. Appl. Cryst. **15**, 567–569 (1982).

[4]B. B. Mandelbrot, *The Fractal Geometry of Nature* (Freeman, San Francisco, 1982).

[5a]P. Pfeifer and D. Avnir, J. Chem. Phys. **79**, 3558–3565 (1983).

[5b]D. Avnir, D. Farin, and P. Pfeifer, J. Chem. Phys. **79**, 3566–3571 (1983).

[6]D. Avnir, D. Farin, and P. Pfeifer, Nature (London) **308**, 261–263 (1984).

[7]Ref. 2, pp. 3 and 4.

[8]Ref. 2, pp. 78–82.

[9]Ref. 2, Eqs. (106) and (107), pp. 78 and 79.

[10]Ref. 2, pp. 12–16.

[11]Ref. 5a, p. 3559.

[12]Ref. 2, p. 78.

[13]A. Erdélyi, *Asymptotic Expansions* (Dover, New York, 1956), pp. 49 and 50.

[14]M. Kalliat, C. Y. Kwak, and P. W. Schmidt, in *New Approaches in Coal Chemistry,* edited by B. D. Blaustein, B. C. Bockrath, and S. Friedman, ACS Symposium Series No. 169 (American Chemical Society, Washington, D.C. 1981), pp. 3–22.

[15]H. Kaiser and J. S. Gethner, Proceedings of the 1983 International Conference on Coal Science, Pittsburgh, Pa., August 1983, pp. 301–303.

[16]J. S. Gethner, private communication.

[17]S. K. Sinha, Y. Freltoft, and J. Kjems, Proceedings of the International Conference on Kinetics of Aggregation and Gelation, Athens, Ga., 1984 (to be published).

[18]D. W. Schaefer and K. D. Keefer, to be published.

[19]Ref. 5a, Sec. VII, p. 3563.

Reprinted with permission from *Nature* **308**, B.B. Mandelbrot, D.E. Passoja, and A.J. Paullay, "Fractal Character of Fracture Surfaces of Metals," pp. 721–722, ©1984 Macmillan Magazines Ltd., London.

Fractal character of fracture surfaces of metals

Benoit B. Mandelbrot*, Dann E. Passoja† & Alvin J. Paullay‡

* IBM Research Center, Yorktown Heights, New York 10598, USA
† Union Carbide Corporation, Tarrytown, New York 10591, USA
‡ Columbia University New York, Bronx Community College, Bronx, New York 10753, USA

When a piece of metal is fractured either by tensile or impact loading (pulling or hitting), the fracture surface that is formed is rough and irregular. Its shape is affected by the metal's microstructure (such as grains, inclusions and precipitates, whose characteristic length is large relative to the atomic scale), as well as by 'macrostructural' influences (such as the size, the shape of the specimen, and the notch from which the fracture begins). However, repeated observation at various magnifications also reveals a variety of additional structures that fall between the 'micro' and the 'macro' and have not yet been described satisfactorily in a systematic manner. The experiments reported here reveal the existence of broad and clearly distinct zone of intermediate scales in which the structure is modelled very well by a fractal surface. A new method, slit island analysis, is introduced to estimate the basic quantity called the fractal dimension, *D*. The estimate is shown to agree with the value obtained by fracture profile analysis, a spectral method. Finally, *D* is shown to be a measure of toughness in metals.

Fractals are the concern of a new geometry[1], whose primary object is to describe the great variety of natural structures that are irregular, rough or fragmented, having irregularities of various sizes that bear a special 'scaling' relationship to one another. In very loose terms, they seem to fall into a regular hierarchy in which each level is an up-sized or down-sized version of the levels below or above it. Fractal geometry characterizes the scaling structure of a surface by a number D, called the fractal dimension, that can range from 2, when the surface is smooth, up to 3. The fractal dimensional increment of a surface is $D-2$. As it increases from 0 to 1, the irregularities become increasingly predominant and the notion of overall shape of the surface becomes progressively less meaningful. Similarly, the fractal dimension of a curve is at least 1, the fractal dimensional increment being $D-1$.

The term 'fractal' was chosen in explicit cognizance of the fact that the irregularities found in fractal sets are often strikingly reminiscent of fracture surfaces in metals (though not, for example in glass). However, metal fractures are only extremely crinkly (down to the limits of their microstructural size range), while fractals are infinitely crinkly. Hence, it is not possible to say that metal fracture is strictly a fractal. Nevertheless, metals resemble a fractal so closely that it makes good sense to use a metal for modelling a fractal. Our experiments in metal fracture show that D is very well defined for all the specimens examined, and takes on the same value for different specimens of the same metal having similar thermomechanical treatments.

Recent studies[2,3] conclude that a metal fracture's spectrum is indicative of a fractal structure. A recent review[4] also indicates that fractals might model metal fracture surfaces, and reports exploratory studies. The more refined tools that we use to analyse metal fracture surfaces make a stronger case and add to the existing tools of fractographic analysis[4,5].

Our first series of tests is exemplified by Fig. 1. A fractured steel specimen was plated with electroless nickel and mounted in an epoxy mount by vacuum impregnation in order to ensure edge retention. The specimen was then polished parallel to the plane of fracture. 'Islands' of steel surrounded by nickel appeared which, on subsequent polishing, grew and merged. We propose to call these structures slit islands. The islands

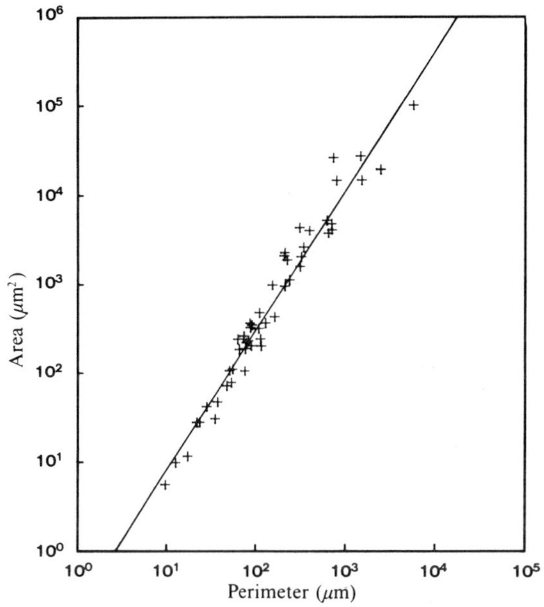

Fig. 1 Fractal area–perimeter relationship for slit islands. 300-Grade Maraging steel. Ruler = 1.5625 μm; fractal dimensional increment = 1.28.

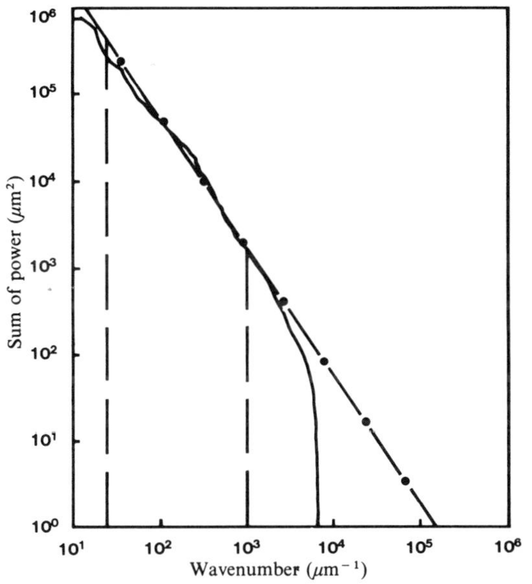

Fig. 2 Cumulative spectrum for vertical cuts. 300-Grade Maraging steel. Slope = 1.4878; fractal dimensional increment = 1.26. Average of five profiles.

contain 'lakes within islands' and 'islands within lakes'; we include the former and neglect the latter.

The slit island 'coastlines', being curves, are much easier to investigate than the surfaces themselves. Yet we have found that the perimeter–area relationship of fractal dimensional analysis (ref. 1, Chap. 12), if applied to these coastlines, reveals many essential facts about the surface. We propose that the resulting method of analysis, which is entirely new and very powerful, be called slit island analysis. When islands are derived from an initial fractal surface of dimension D by sectioning with a plane, their coastlines are of fractal dimension $D'=D-1$. Thus, the fractal dimensional increments $D-2$ and $D'-1$ are equal. The theory of fractals suggests that the areas and

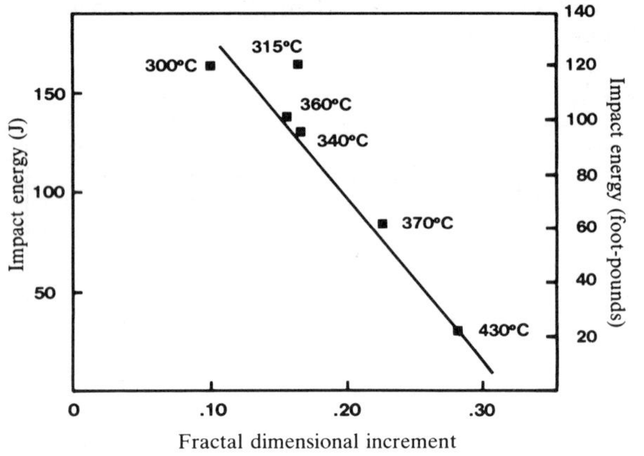

Fig. 3 Impact energy versus fractal dimensional increment. 300-Grade Maraging steel, room temperature, Charpy impact.

perimeters of these islands should be measured in the same way, and that one should trace the graph of log(perimeter) versus log(area). In a fractal, this graph is rectilinear of slope D'. Figure 1 shows that such is indeed the case for fracture surfaces.

Our second series of tests of fractal structure which complements the first, uses fracture profile analysis based on Fourier analysis[3] (Fig. 2). Here, the plated and processed fracture was sectioned perpendicular to the fracture surface to expose it in profile. As usual, the profile's sample spectra exhibit wide oscillations. Some are statistical artefacts, but others reflect fundamental lengths of the microstructure and their accompanying higher order harmonics. Statistical fluctuations must be averaged out and the fluctuations due to the microstructure should be prevented from overly affecting the analysis. To achieve both goals, we averaged five spectra taken from serial sections, and integrated them from high frequency to low. Now, the fractal hypothesis requires the integrated spectrum to take the form $k^{-B'}$, with $B'=B-1=6-2D$. The fractal character of the surface was tested by plotting the data on doubly logarithmic coordinates to see if the curve has a large straight central portion of absolute slope B'. The plots (Fig. 2) are indeed straight but over narrower ranges than the area–perimeter plots. Thus, the fracture surface is again inferred to be fractal, and its fractal dimension is estimated by $D=3-B'/2$.

When a method of analysis yields a nearly straight graph, it is tempting to measure the 'local' slope of this curve. The goal would be to identify the harmonics of the microstructure and other features that (contrary to the overall fractal behaviour) are defined by well-defined length scales. However, as the local slope is severely affected by statistical fluctuations, attempts to interpret its variations are useless. Similarly, the 'spectrum of fractal dimension' described in ref. 5 mostly reflects the underlying noise.

Careful analysis of some of our data indicates that the central region of the log–log plots splits into two distinct subregions characterized by different values for D. Such a juxtaposition of different fractal zones is very familiar in other applications of fractals. The scales of the clear-cut cross-over points between zones tend to be significant with respect to structure and deserve further investigation. For other metals, the diagrams are even more complex.

To relate D to known metallurgical quantities, we took a series of identical 300 grade Maraging steel Charpy impact specimens, and heat-treated them at different temperatures. Figure 3 shows that the value of D decreases smoothly with an increase of the impact energy as measured by a standard Charpy impact test. This relationship must reflect the changes in the microstructure that occur during ageing.

In our view, a fracture that is transgranular (that is, runs through the grains) involves an atypical form of the notion of percolation. For example, during fracture of a ductile material, any voids forming around inclusions increase in size and coalesce into void sheets; these ultimately form the fracture surface. If the growth of a void were independent of its neighbours or its position in the specimen, we would be faced with a process physicists call percolation. Its applicability here is associated with load sharing among the tensile ligaments that remain; void sheet coalescence is merely percolation of a crack in the microstructure leading to final fracture. Percolation would imply that the fracture's fractal dimension takes some universal value independent of the material, and dependent only on the properties of space. But percolation is a very crude model here. As soon as the initial void growth has coalesced locally into small sheets, the strains on the supporting ligaments increase, and surrounding voids grow at a rate that varies with their position in the specimen. Certainly, spatial variability associated with the microstructure is dependent on the microstructure and the resulting process differs from the usual percolation. Were this view to be confirmed by further experimentation, the study of fracture could benefit from techniques used in the study of phase transitions in statistical physics.

By showing linear relationships among all available data while covering such a broad range of sizes, Figs 1 and 2 are almost unique in metallurgy. Their linearity demonstrates that the fracture surfaces present in several grades of steel produced by impact or uniaxial tensile loading are fractal.

The metallographic assistance of H. I. Kohlmorgen is gratefully acknowledged. Comments by J. A. Given helped clarify this text.

Received 4 November 1983; accepted 18 January 1984.

1. Mandelbrot, B. B. *The Fractal Geometry of Nature* (Freeman, New York, 1982); also *Les objets fractals* (Flammarion, Paris, 1975, 1984) and *Fractals; Form, Chance and Dimension* (Freeman, New York, 1977).
2. Passoja, D. E. & Psioda, J. A. in *Fractography and Materials Science* (eds Gilbertson, L. N. & Zipp, R. D.) 335–386 (American Society for Testing Materials, Philadelphia, 1981).
3. Passoja, D. E. & Amborski, D. J. in *Microstruct. Sci.* **6,** 143–148 (1978).
4. Gilbertson, L. N. & Zipp, R. D. (eds) *Fractography and Materials Science*, ASTM STP 733 (American Society for Testing Materials, Philadelphia, 1981).
5. Coster, M. & Chermant, J. L. *Int. Metals Rev.* **28,** 234–238 (1983).

Printed in Great Britain by Macmillan Production Ltd., Basingstoke, Hampshire

Reprinted with permission from *Nature* **308**, D. Avnir, D. Farin, and P. Pfeifer, "Molecular Fractal Surfaces," pp. 261–263,

Molecular fractal surfaces

David Avnir & Dina Farin

Department of Organic Chemistry, The Hebrew University of Jerusalem, Jerusalem 91904, Israel

Peter Pfeifer

Fakultät für Chemie, Universität Bielefeld, D-4800 Bielefeld, FRG

In this letter we report that, at the molecular-size range, the surfaces of most materials are fractals, that is, at this range, surface geometric irregularities and defects are characteristically self-similar upon variations of resolution. The whole range of fractal dimension[1], $2 \leq D < 3$, is found in the many examples presented. Two representative examples, namely adsorption of polystyrene on alumina and adsorption of krypton on dolomite are discussed in some detail. Our findings suggest a simple solution to the problem of quantifying the degree of surface irregularity[2,3] at a resolution which is of relevance to many aspects of surface science. The results provide a general explanation for phenomenological links between various surface parameters, and we derive a set of equations of use in predicting surface variables.

Surface-science has in the past tended to deal with the problem of description of geometrically irregular surfaces, such as those found in oxide gels, carbons, catalysts, metals[4], by treating irregularity as a deviation from well-defined Eucleidean shapes. An entirely different approach is possible, if the surface is (statistically) invariant over a certain range of scale transformations—that is, if the surface is self-similar upon changes in resolution power. For such surfaces, the degree of irregularity can be given by one number: the fractal dimension[1], $2 \leq D < 3$, of the surface. The higher the D value, the more wiggly and space filling the surface is (see below). The obvious question is then, is self-similarity common in nature so as to make the fractal-approach useful? The similarity of computer-generated fractal-objects to real objects led Mandelbrot[1] to advocate strongly a positive answer to that question; and in recent years several reports have appeared[1,5-7] in which boundary lines of natural objects ('coast lines') have been found to be fractals ($1 \leq D < 2$). Similarly, natural landscapes were reported to be fractals[8,9]. On the molecular scale, the only direct experimental fractal measurement is, to our best knowledge, the analysis of back-bone 'wiggliness' in proteins[10].

Here we present an extensive compilation suggesting that the fractal approach is indeed of value. Moreover, in contrast to most previous fractal analyses, which probed the $1 \leq D < 2$ range of fractal dimensions, and could only indirectly provide the surface fractal dimensions, our methods directly probe the $2 \leq D < 3$ range.

The fractal theory of surface science we have developed[2,3] is an extension of the basic relation[1] which describes self-similarity:

$$n \propto r^{-D} \qquad (1)$$

where n is the number of yardsticks of size r necessary to cover (in order to measure) the length or the surface area of the fractal object. Note that for, say, flat smooth surfaces, equation (1) becomes the familiar $n \propto r^{-2}$, and that volume-filling is $n \propto r^{-3}$. In our application, the yardsticks are molecules. The equations used are shown in Table 1.

Following our finding that silica gel has a fractal surface[2], we found the surfaces of some other common materials to be fractals (ref 11 and Table 2). Further analysis of published data (Table 2) lead us to conclude that at the molecular range, the surfaces of most materials are fractals. Well over 75% of the analysed surfaces fell into this category. Table 2 emphasizes the wide diversity of materials which possess a fractal surface, and the ability of our method to probe the irregularity of hindered areas within porous materials. The complete range ($2 \leq D < 3$) of fractal dimensions seems to be found in nature: low $D = 2.0$ values, indicate regularity and smoothness (for example, Aerosil), D values deviating slightly from 2 are indicative of a surface with slight defects (2.07 for graphite), intermediate D values represent a fairly irregular surface (some types of charcoal), and D values close to 3 indicate extreme irregularity of the surface (silica gel). A wide range of self similarity can be detected when equations (7) or (8) are used, or when the yardsticks are polymer fractions. The smaller ranges found in Table 2 are artificially limited by the given range of yardstick sizes. The processes by which these fractal surfaces were formed probably operated over a much wider scale, as can be judged by comparison of ranges of similar materials, such as charcoals. In any event, even the limited-range examples carry information of interest, since they cover the size of all small organic molecules (C_1 to C_{10}).

Two examples taken from Table 2 serve to illustrate the nature of those fractal surfaces. The first is the surface of porous alumina, (Alcoa grade F-20) studied by Burns and Carpenter[12]. They adsorbed monolayers of nitrogen and four spherically coiled polystyrene fractions, having radii of gyration in the range 73–379 Å. The decrease in the resulting apparent surface area, as a function of yardstick size, is faster than $n \propto r^{-2}$ (for smooth surfaces), and obeys equation (4), as depicted in Fig. 1. From the slope of the straight line, the fractal dimension of the surface

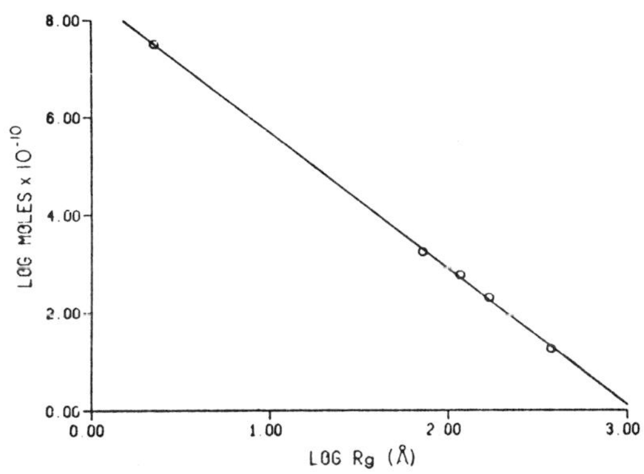

Fig. 1 Aluminium oxide, $D = 2.79 \pm 0.03$. Monolayer amount of polystyrene molecules and nitrogen (mol $\times 10^{-10}$ per g), as a function of the radius of gyration R_g (Å).

Table 1 Equations derived for determination of fractal dimension

Equation no.		Comments
(1)	$n \propto r^{-D}$	See text
(2)	$n \propto \sigma^{-D/2}$	—
(3)	$A \propto \sigma^{(2-D)/2}$	—
(4)	$n \propto R_g^{-D}$	—
(5)	$n \propto v^{-D+1}$	Holds only for series of homologous linear absorbates lying flat on surface
(6)	$V_m \propto \sigma^{-D/2}$	—
(7)	$A \propto R^{D-3}$	—
(8)	$n \propto R^{D-3}$	—
(9)	$\sigma_0 \leq \sigma \leq \sigma_0 (R_{max}/R_{min})^2$	—

In these equations, n is the monolayer coverage (for example, moles of adsorbate per gram adsorbent; σ is the effective cross-sectional area of adsorbate; A is the apparent surface area ($m^2\,g^{-1}$); R_g is the radius of gyration; v is the molar volume; V_m is the monolayer coverage as expressed by the STP volume in the BET method; R, R_{max} and R_{min} are values for particle radius; σ_0 is the cross-sectional area of adsorbate, as used for equations (7) and (8).

Table 2 Fractal dimensions in a range of materials

Fractal dimension*	Found in	At the range (Å^2)†	Adsorbates	Method of D determination‡	Ref.§
High					
2.91 ± 0.02	Upper Columbus dolomitic rock from Bellevue, Ohio	20–47,000	Kr	(7)	13
2.97 ± 0.01	Goodland high calcium rock, from Idabel, Okla.	20–47,000	Kr	(7)	13
2.88 ± 0.02	Granitic rock from SHOAL nuclear test site, Nevada	16–16,500	N_2	(7)	14
2.73 ± 0.05	Igneous rock sample from SHOAL site	14–14,300	Ar	(7)	14
2.71 ± 0.14	Granular activated carbon–Tsurumi HC-8, from coconut shell	(16–37)	N_2, 12 various organic molecules	(2)	15
2.80 ± 0.16	Granular activated carbon—Fujisawa B-CG, from coconut shell	(16–37)	N_2, 12 various organic molecules	(2)	15
2.90 ± 0.01	Carbonate rock from groundwater test well, Yucca Flat, Nevada	16–16,500	N_2	(7)	14
2.92 ± 0.02	Soil (kaolinite, trace halloysite)	150–16,500	Malachite green	(8)	16
2.94 ± 0.04	Porous silicic acid	(16–34)	N_2, MeOH, EtOH, PrOH, iso-PrOH, BuOH	(3)	2
2.79 ± 0.03	Activated alumina grade F-20 (Alcoa Corp.)	16–45,100	Polystyrene fractions, N_2	(4)	12
2.78 ± 0.21	Charcoal (BDH) of animal origin	1,400–180,000	Styrene-methyl-methacrylate random copolymers	(4)	11
2.67 ± 0.16	Porous coconut charcoal (Standard Chemical Co., Montreal)	(16–47)	N_2, C_2H_2, C_2H_4, CH_4, C_2H_6, C_3H_8, n-C_4H_{10}, iso-C_4H_4	(2)	11
Medium					
2.57 ± 0.04	Porous α-FeOOH pigment for magnetic tapes	16–980	N_2	(7)	17
2.35 ± 0.11	Crushed Corning 0010 lead glass	21–14,900	Hexadecyl-l-^{14}C-trimethyl ammonium bromide	(7)	11
2.52 ± 0.07	Coal mine dust from Western Pennsylvania	16–180	N_2	(7)	18
2.33 ± 0.08	Coal mine dust from Western Pennsylvania	16–270	N_2	(7)	18
2.25 ± 0.09	Carbon black	(16–71)	N_2, benzene, naphthalene, anthracene, phenantrene	(2)	11
2.54 ± 0.12	Slightly porous coconut charcoal (Standard Chemical Co., Montreal)	(16–47)	N_2, C_2H_2, C_2H_4, CH_4, C_2H_6, C_3H_8, C_4H_{10}, iso-C_4H_{10}	(2)	11
2.30 ± 0.07	Slightly porous coconut charcoal (Standard Chemical Co., Montreal)	(16–47)	N_2, C_2H_2, C_2H_4, CH_4, C_2H_6, C_3H_8, C_4H_{10}, iso-C_4H_{10}	(2)	11
2.63 ± 0.03	Mosheim high calcium from Stephens City, Va.	20–47,000	Kr	(7)	13
2.58 ± 0.01	Niagara (Guelph) dolomite from Woodville, Ohio	20–47,000	Kr	(7)	13
2.46 ± 0.11	Glassy melted rock from Rainier nuclear zone, Nevada	14–14,300	Ar	(7)	14
2.29 ± 0.06	Soil (mainly feldspar quartz and limonite)	150–21,800	Malachite green	(8)	16
Low					
2.02 ± 0.06	Aerosil—nonporous fumed silica (Degussa)	16–529	N_2	(7)	11
2.15 ± 0.06	Snowit—ground fine Belgian quartz glass of high purity	16–10,600	N_2	(7)	19
2.14 ± 0.06	Madagascar quartz from Thermal Syndicate	16–1,850	N_2	(7)	19
1.95 ± 0.04	Periclase—electrically fused and crushed magnesite	16–720	N_2	(7)	20
2.02 ± 0.05	Synthetic faujasite ($Na_2O \cdot Al_2O_3 \cdot 2{,}67SiO_2 \cdot mH_2O$) (Linde Air Products)	(16–68)	N_2, n-alkanes (C_1, C_{3-7})	(5)	11
2.07 ± 0.01	Graphite—Vulcan 3G (2700) (National Physical Lab., Teddington, UK)	(16–178)	N_2, n-alkanes (C_{22}, C_{28}, C_{32})	(2)	11
2.04 ± 0.16	Graphon—partially graphitized carbon black formed by heating MPC black to 3,200 °C	(15–41)	N_2, MeOH, EtCl, benzene, cyclohexane	(6)	11
2.13 ± 0.16	Graphon—graphitized carbon black (Cabot Corp.)	1,400–180,000	Styrene-methyl-methacrylate random copolymers	(4)	11
2.04 ± 0.04	Active, (nonporous), coconut charcoal (Standard Chemical Co., Montreal)	(16–47)	N_2, C_2H_2, C_2H_4, CH_4, C_2H_6, C_3H_8, n-C_4H_{10}, iso-C_4H_{10}	(2)	11
1.97 ± 0.02	Active, (nonporous), coconut charcoal (Standard Chemical Co., Montreal)	(16–47)	N_2, C_2H_2, C_2H_4, CH_4, C_2H_6, C_3H_8, n-C_4H_{10}, iso-C_4H_{10}	(2)	
2.16 ± 0.04	Iceland spar, massive calcite from Chihuahua, Mexico	20–47,000	Kr	(7)	

* Error estimation indicates the least-square analysis of best fit to a straight line.

† Values in parentheses give the range of yardstick sizes, when the full range cannot be estimated directly either from equation (9) or from polymer adsorption. For nitrogen, the standard used value of 16.2 Å^2 is taken.

‡ Numbers refer to equation nos in Table 1.

§ All experimental data have been published previously: refs 2, 11 contain details on data and its analyses; refs 12–20 are data sources only.

of the studied alumina is 2.79 ± 0.03. The range of surface self-similarity found for this example is high, covering three orders of magnitudes. This observation was made possible by the remarkable experimental result that monolayer values obtained by different methods (Brunauer, Emmett and Teller (BET) surface area test at cryogenic temperatures for the small nitrogen; adsorption from cyclohexane solution for the large polymer yardsticks) fit closely the same straight line—that is, the geometric factor rather than the nature of physisorption, dominates. The self-similarity of this surface is further demonstrated by the fact that whereas the large polymers can penetrate and probe only the widest pores (and the outer surface), nitrogen probes much of the pore network that is hidden from the polymer.

The second example involved a different method of D-determination, the analysis of change in apparent surface area as a function of particle size[3] (equations (7) and (8)). The method has several advantages: (1) it is easier to obtain a wide range

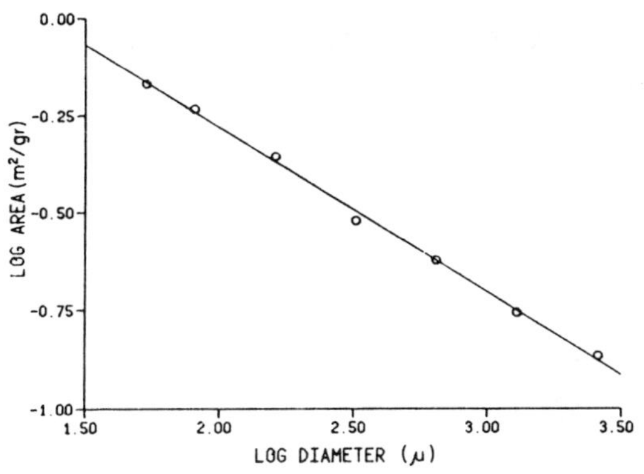

Fig. 2 Niagara dolomite, $D=2.58\pm0.01$. BET (Kr) surface area (m^2 per g) shown as a function of particle diameter (μm).

of particle sizes than to use a wide range of molecular yardsticks, so that, a wide range of self-similarity may be detected; (2) adsorption of only one type of molecule is necessary; (3), the measurement scale in this approach is the particle-radius so that an exact knowledge of the effective cross section of the probe molecule is necessary only for estimation of the range of self-similarity (equation (9)). Figure 2 summarizes the example, in which Love and Whittaker measured the surface area of limestones from various sources by krypton adsorption[13]. Niagara (Guelph) dolomite from Woodville, Ohio, was crushed and sieved into seven fractions ranging from 53.5 to 2,605 μm in size, and from equation (7) we find $D=2.58\pm0.01$. All the other rock-surfaces studied by Love and Whittaker were found to be fractals (see all Kr entries in Table 2).

Thus the geometry of materials has conventionally been classified in terms of types of crystallinity or amorphousness; we now propose fractality as a new viewpoint in material geometry classification. Since it is difficult to conceive of any dynamic interfacial process that is independent of surface irregularity or surface defects, the importance of such classification seems obvious. Catalysis is one phenomenon which is governed by such surface characteristic and this is now being studied.

This work was supported by the Fritz Haber Research Center for Molecular Dynamics, Jerusalem, the Israel–US Binational Foundation (D.A.) and by the Ben-Gurion Foundation (to D.F.). D.A. is an M. Richter Fellow.

Received 23 August; accepted 21 December 1983.

1. Mandelbrot, B. B. *The Fractal Geometry of Nature* (Freeman, San Francisco, 1982).
2. Avnir, D. & Pfeifer, P. *Nouv. J. Chim.* **7**, 71–72 (1983).
3. Pfeifer, P. & Avnir, D. *J. chem. Phys.* **79**, 3558–3565 (1983).
4. Beddow, J. K. & Meloy, T. P. (eds) *Advanced Particulate Morphology* (CRC, Florida, 1977).
5. Lovejoy, S. *Science* **216**, 185–187 (1982).
6. Paumgartner, D., Losa, G. & Weibel, E. R., *J. Microsc.* **121**, 51–63 (1981).
7. Kaye, B. H. *Direct Characterization of Fine Particles* (Wiley, New York, 1981).
8. Berry, M. V. & Hannay, J. H. *Nature* **273**, 573 (1978).
9. Burrough, P. A. *Nature* **294**, 240–242 (1981).
10. Allen, J. P., Calvin, J. T., Stinson, D. G., Flynn, C. P. & Stapelton, H. J. *Biophys. J.* **38**, 299–310 (1982).
11. Avnir, D., Farin, D. & Pfeifer, P. *J. chem. Phys.***79**, 3566–3571 (1983).
12. Burns, H. Jr & Carpenter, D. K. *Macromolecules* **1**, 384–390 (1968).
13. Love, K. S. & Whittaker, C. W. *Agr. Food. Chem.* **2**, 1268–1272 (1954).
14. Carnahan, C. L., Castagnola, D. C. & Smith, M. E. U.S. Atomic Energy Comm. NVO-1229-98 (1968).
15. Urano, K., Omori, S. & Yamamoto, E. *Env. Sci. Tech.* **16**, 10–14 (1982).
16. Cohen, W. H. & Knight, B. H. *J. Soc. Chem. Ind.* **66**, 357–364 (1947).
17. Suzuki, S. *et al. Ferrites*, 556–560 (Tokyo, 1980).
18. Corn, M. *et al. Am. ind. Hygiene J.* **34**, 279–285 (1973).
19. Cartwright, J., Wheatley, K. & Sing, K. S. W. *J. appl. Chem.* **8**, 259–264 (1958).
20. Girgis, B. S. & Girgis, L. G. *Powder Tech.* **7**, 85–91 (1973).

Reprinted with permission from *Optica Acta* **33**, M.V. Berry and I.C. Percival, "Optics of Fractal Clusters such as Smoke," pp. 577–591, ©1986 Taylor & Francis Ltd., London.

Optics of fractal clusters such as smoke

M. V. BERRY

H. H. Wills Physics Laboratory, Tyndall Avenue, Bristol BS8 1TL, England

and I. C. PERCIVAL

School of Mathematical Sciences, Queen Mary College, University of London, Mile End Road, London E14NS, England

(*Received 30 October 1985*)

Abstract. Scalar or vector light of wavelength $2\pi/k$ strikes N small refracting and absorbing spherules, each of radius a, which have coagulated into a sparse random cluster with fractal dimension D (for smoke, $D \approx 1{\cdot}78$). It is assumed that $ka \ll 1$ but the cluster size $R = aN^{1/D}$ may be larger than the wavelength. Using a mean field theory it is shown that multiple scattering is negligible for all N if $D<2$, and becomes important when $N \sim (ka)^{-D/(D-2)}$ if $D>2$. Cross-sections are calculated as functions of N, D and the complex refractive index of the spherules. If $D<2$ the scattering cross-section per spherule rises with N and saturates when $kR \gg 1$, at a value exceeding that of an isolated spherule by a factor of order $(ka)^{-D}$; if $D>2$ the same quantity increases as $N^{1-2/D}/(ka)^2$. For $D<2$ the absorption cross-section is N times that of a solitary spherule. These results are very different from those for spherules coagulating into compact solid spheres ($D=3$), and are important for the optics of sooty smoke, with $D \approx 1{\cdot}78$, which is used as an example throughout.

1. Introduction

Smoke is formed as small soot spherules [1] which stick together when they touch and thereby coagulate. Further aggregation leads to sparse random clusters with a fractal dimension of about 1·78 [2, 3]. This fractality has not been included in studies of the optics of smoke. The omission is important for the scattering and extinction of radiation by smoke lofted into the atmosphere by large fires, such as those produced by multiple nuclear explosions [4–7]. Our purpose here is to present a theory of the optics of smoke in which the importance of the fractal structure is fully brought out. Implications for the nuclear winter predictions will be discussed in a separate publication.

The central simplifying principle will be that the wavelength λ of the illuminating radiation, which is about $0{\cdot}5\,\mu$ for the visible and $10\,\mu$ in the infrared, greatly exceeds the spherule radius a, which is about $0{\cdot}005\,\mu$–$0{\cdot}05\,\mu$ for sooty smoke. In other words, the observation scale is much bigger than the inner fractal scale. However, the size R of a cluster (that is, the outer fractal scale) increases with aggregation (§ 2) as the number of spherules in a cluster increases, from a (when $N=1$) to values which may considerably exceed λ (when N is several thousand). We shall calculate the optical cross-sections as functions of N and show that the optics of clusters with $R \ll \lambda$ and $R \gg \lambda$ can be very different. Moreover the differences depend strongly on whether the fractal dimension D is less than 2 or greater than 2.

The complications of vector wave theory obscure the essential ideas and so this case will be treated last (§§ 6–8), after the scalar theory (§§ 3–5) (which applies to scattering of scalar sound waves and to neutron optics, and also as a good approximation for light).

2. Cluster statistics

Consider the ensemble consisting of all those clusters containing N spherules. The *size* R of these N-clusters is defined as the r.m.s. distance between spherules, that is, $\sqrt{2}$ times the radius of gyration. For clusters with fractal dimension D (which satisfies $1 \leqslant D \leqslant 3$ because each cluster is connected) the size is related to the spherule radius a by

$$R = aN^{1/D}. \tag{1}$$

(Strictly speaking we should write $R = \alpha a N^{1/D}$, but $\alpha = 1$ is a close approximation: for smoke ($D = 1{\cdot}78$), the simulations of Mountain and Mulholland [3] give $\alpha = 0{\cdot}96$; for a linear array ($D = 1$), $\alpha = (\frac{2}{3})^{1/2} = 0{\cdot}82$; for a random walk of touching spherules ($D = 2$), $\alpha = 2/\sqrt{3} = 1{\cdot}15$; for a close-packed 'sphere' of spherules ($D = 3$), $\alpha = 1{\cdot}11$.) Equation (1) embodies the sparseness of fractal clusters, because the number density of spherules decreases with aggregation as

$$\rho \sim \frac{N}{R^3} = \frac{1}{a^3 N^{3/D - 1}}. \tag{2}$$

Fractality implies not only sparseness but also self-similarity and strong clustering. When applied to the optically important probability density $P(\mathbf{r})$ for finding a randomly chosen pair of spherules separated by a vector $\mathbf{r}_j - \mathbf{r}_l = \mathbf{r}$, self-similarity, together with isotropy, means that

$$P(\mathbf{r}) = \frac{1}{4\pi R^3} p(r/R) \tag{3}$$

where

$$\int_0^\infty \mathrm{d}x\, x^2 p(x) = 1 \quad \left(\text{normalization, and} \quad \int_0^\infty \mathrm{d}x\, x^4 p(x) = 1 \quad \text{size } R\right). \tag{4}$$

Strong clustering means that the number of particles in a sphere of radius $\delta (\ll R)$ centred on a spherule is proportional to δ^D, so that

$$p(x) \to \frac{1}{x^{3-D}} \quad \text{as} \quad x \to 0. \tag{5}$$

For detailed calculations we will take the convenient form

$$p(x) = \frac{A}{x^{3-D}} \exp(-bx) \tag{6}$$

whose constants, determined by (4), are

$$A = [(D+1)D]^{D/2}/\Gamma(D), \quad b = [(D+1)D]^{1/2}. \tag{7}$$

When $D=1{\cdot}78$, $A=4{\cdot}48$, $b=2{\cdot}22$. For comparison, the exact interspherule distance probability for a random walk ($D=2$) has

$$p(x)=\frac{8}{(2\pi)^{1/2}}\left((1+x^2)\int_1^\infty \mathrm{d}v \exp\left(-\tfrac{1}{2}x^2v^2\right)-\exp\left(-\tfrac{1}{2}x^2\right)\right). \tag{8}$$

This is derived by averaging the Gaussian probability distribution for $\mathbf{r}_j-\mathbf{r}_l$ (which has variance proportional to $|j-l|$ if j and l label positions along the walk) over the N values of j and l. It has the limiting forms

$$p(x) \begin{matrix} \nearrow & 4/x, & \text{as} \quad x\to 0 \\ \searrow & \dfrac{16\exp(-x^2/2)}{(2\pi)^{1/2}x^4}, & \text{as} \quad x\to\infty. \end{matrix} \tag{9}$$

We emphasize that provided $p(x)$ satisfies the conditions (4) and (5) (and also decays faster than x^{-1} as $x\to\infty$) its detailed form will not affect any of our conclusions.

Scaling breaks down at the inner fractal scale and we should take $p(x)=0$ for $x<2a/R$. However, for all our optical averages over pairs of spherules, integrals over $p(x)$ will converge with the limiting form (5). Therefore the small-x cut off would introduce a negligible correction and we henceforth ignore it.

3. Scalar waves: scattering equations

Each spherule will be considered to have the same refractive index

$$\mu=\mu_1+\mathrm{i}\mu_2. \tag{10}$$

For smoke, the transmission and absorption indices μ_1 and μ_2 depend on wavelength and on its physical and chemical nature. Representative values for soot (estimated from data of Chylek *et al.* [1]) are

$$\left.\begin{matrix} \mu=1{\cdot}75+0{\cdot}3\mathrm{i}, & \text{visible} \\ \mu=3+\mathrm{i}, & \text{infrared}. \end{matrix}\right\} \tag{11}$$

Scalar waves $\psi(r)$ satisfy

$$\nabla^2\psi(\mathbf{r})+k^2\mu^2(\mathbf{r})\psi(\mathbf{r})=0 \tag{12}$$

where $k\equiv 2\pi/\lambda$, $\mu(\mathbf{r})=\mu$ inside a spherule and zero outside, and ψ and $\nabla\psi$ are continuous. The incident wave

$$\psi_i=\exp(\mathrm{i}\mathbf{k}_\mathrm{i}\cdot\mathbf{r}) \tag{13}$$

interacts with the N-cluster and produces the scattered wave, whose asymptotic form is

$$\psi_\mathrm{s}=\frac{\exp(\mathrm{i}kr)}{r}f(\mathbf{k}_\mathrm{s},\mathbf{k}_\mathrm{i}) \tag{14}$$

when $\mathbf{k}_\mathrm{s}\equiv k\mathbf{r}/r$ and f is the scattering amplitude. We seek to calculate the ensemble averages of the differential cross-section

$$\mathrm{d}\sigma(\mathbf{k}_\mathrm{s},\mathbf{k}_\mathrm{i})/\mathrm{d}\Omega\equiv|f(\mathbf{k}_\mathrm{s},\mathbf{k}_\mathrm{i})|^2, \tag{15}$$

the scattering cross-section

$$\sigma_s(\mathbf{k}_i) \equiv \int\int \mathrm{d}\Omega_s \frac{\mathrm{d}\sigma}{\mathrm{d}\Omega}(\mathbf{k}_s, \mathbf{k}_i) \tag{16}$$

(where Ω_s is the direction of $\mathbf{k}_s$), the extinction cross-section

$$\sigma_e(\mathbf{k}_i) = (4\pi/k)\,\mathrm{Im}\, f(\mathbf{k}_i, \mathbf{k}_i) \tag{17}$$

and the absorption cross-section

$$\sigma_a(\mathbf{k}_i) = \sigma_e(\mathbf{k}_i) - \sigma_s(\mathbf{k}_i). \tag{18}$$

The fundamental approximation is $ka \ll 1$ (small spherules); thus each spherule scatters isotropically and weakly. At this stage the Born approximation is the simplest approximation for individual scatterings but it is not adequate for our purposes because its failure to satisfy the optical theorem for real μ implies a false absorption in addition to the true absorption when μ is complex. Instead we write the scattering amplitude for a single spherule as

$$f_1 \equiv \tau/k \quad \text{with} \quad \tau \equiv \sin\eta \exp \mathrm{i}\eta \tag{19}$$

where the s-wave phase shift η, obtained by solving (12) [8], is approximately

$$\eta \approx \tfrac{1}{3}(ka)^3(\mu^2 - 1). \tag{20}$$

For real μ, η is real and the identity $|\tau|^2 = \mathrm{Im}\,\tau$ guarantees that the optical theorem holds. For complex μ, for example smoke, the single-spherule cross-sections are

$$\sigma_s^{(1)} \approx \pi a^2 \tfrac{4}{9}(ka)^4 |\mu^2 - 1|^2, \quad \sigma_a^{(1)} \approx \sigma_e^{(1)} \approx \pi a^2 \tfrac{4}{3} ka\, \mathrm{Im}\, \mu^2. \tag{21}$$

For sooty smoke all three cross-sections are much smaller than the spherule's geometrical size, and the absorption greatly exceeds the scattering.

For the cluster, with spherules centred at $\mathbf{r}_1 \ldots \mathbf{r}_N$, the total scattered wave can be written as the sum of wave from each spherule

$$\psi_s = \frac{\tau}{k} \sum_{j=1}^{N} \psi_j \frac{\exp(\mathrm{i}k|\mathbf{r} - \mathbf{r}_j|)}{|\mathbf{r} - \mathbf{r}_j|} \to \frac{\exp(\mathrm{i}kr)}{r} \tau \sum_{j=1}^{N} \psi_j \exp(-\mathrm{i}\mathbf{k}_s \cdot \mathbf{r}_j) \tag{22}$$

where ψ_j is the total wave incident on the jth spherule. This in turn is the sum of the incident wave and those scattered to j from the other spherules:

$$\psi_j = \exp(\mathrm{i}\mathbf{k}_i \cdot \mathbf{r}_j) + \tau \sum_{l \neq j} \psi_l \frac{\exp(\mathrm{i}kr_{lj})}{kr_{lj}} \tag{23}$$

where $\mathbf{r}_{lj} \equiv \mathbf{r}_l - \mathbf{r}_j$. Now we write

$$\psi_j = c_j \exp(\mathrm{i}\mathbf{k}_i \cdot \mathbf{r}_j) \tag{24}$$

where the coefficients c_j incorporate the effects of multiple scattering amongst the spherules. Thus, finally, we have the exact relations

$$f(\mathbf{k}_s, \mathbf{k}_i) = \frac{\tau}{k} \sum_{j=1}^{N} c_j \exp[\mathrm{i}(\mathbf{k}_i - \mathbf{k}_s) \cdot \mathbf{r}_j] \tag{25}$$

where the c_j are determined by

$$c_j = 1 + \tau \sum_{l \neq j} c_l \frac{\exp(ikr_{lj} + \mathbf{k}_i \cdot \mathbf{r}_{lj})}{kr_{lj}}. \tag{26}$$

This equation is the basis of calculations by Jones [9] of the optics of small individual clusters.

4. Scalar waves: multiple scattering estimates

We seek conditions when multiple scattering can be neglected, that is when the $N-1$ terms of the summation in (26) can be neglected, giving $c_j \approx 1$. A crude preliminary bound for multiple scattering can be obtained by iteration, making the successive terms as large as possible by setting all separations r_{lj} equal to the smallest possible value, namely $2a$. This gives

$$c_j \sim (1 - N\tau/2ka)^{-1} \tag{27}$$

so that multiple scattering is certainly negligible if

$$N \ll 2/|\tau| \sim 1/(ka)^2. \tag{28}$$

In terms of the cluster size (1) this condition is

$$kR \ll 1/(ka)^{2/D-1}. \tag{29}$$

When $D<2$, therefore, clusters can be much bigger than the wavelength and still not scatter multiply. When $D>2$ this crude lower bound on N gives a cluster size much smaller than the wavelength and is not useful.

One procedure for getting better estimates might be to study the successive orders of multiple scattering obtained by iterating (26) and averaging each term separately. This is a notoriously difficult procedure because although successive scattering must be from different spherules, later rescatterings from the same spherule do occur, leading to infinite complexity in the scattering topologies which must be summed and averaged (for a survey see [10]). Moreover, the finite order multiple scattering approximations all violate the optical theorem. Another possibility is numerical inversion of the $N \times N$ matrix of coefficients in (26), but while this is feasible for small N [9] we are interested in $N \to \infty$ with ka and D variable.

Here we adopt a mean field approximation, in which all the multiple scattering coefficients c_j have the same value, c, given by averaging the $N(N-1)/2$ matrix elements in (26), so that

$$c = [1 - (N-1)\tau Q]^{-1} \tag{30}$$

where

$$Q \equiv \left\langle \frac{\exp(ikr_{lj}) \cos(\mathbf{k}_i \cdot \mathbf{r}_{lj})}{kr_{lj}} \right\rangle \tag{31}$$

with $\langle\ \rangle$ denoting an average over all interparticle distances r_{lj} in the ensemble of N-clusters. Thus the wave ψ_j incident on each spherule is given by the same multiple of the incident wave. In terms of multiple scattering this is an infinite-order approximation, whose diagrammatic interpretation, for the non-fractal case of the effective refractive index of a statistically homogeneous distribution of scatterers, was discussed by Warner and Gubernatis [11].

A merit of the approximation is that it preserves the optical theorem exactly. From (25), the extinction cross-section is

$$\langle\sigma_e\rangle=\frac{4\pi N}{k^2}\operatorname{Im} c\tau=\frac{4\pi N[\operatorname{Im}\tau+|\tau|^2(N-1)\operatorname{Im}Q]}{k^2|1-(N-1)\tau Q|^2}$$
$$=\frac{4\pi N(\operatorname{Im}\tau+|\tau|^2(N-1)\langle(\sin kr_{lj}/kr_{lj})\cos(\mathbf{k}_i\cdot\mathbf{r}_{lj})\rangle)}{k^2|1-(N-1)\tau Q|^2}. \quad (32)$$

The scattering cross-section is

$$\langle\sigma_s\rangle=\frac{|\tau|^2|c|^2}{k^2}\sum_{j=1}^{N}\sum_{l=1}^{N}\left\langle\int\int d\Omega_s \exp[i(\mathbf{k}_s-\mathbf{k}_i)\cdot\mathbf{r}_{lj}]\right\rangle$$
$$=\frac{4\pi N|\tau|^2(1+(N-1)\langle(\sin kr_{lj}/kr_{lj})\cos(\mathbf{k}_i\cdot\mathbf{r}_{lj})\rangle)}{k^2|1-(N-1)\tau Q|^2}. \quad (33)$$

The absorption cross-section is thus

$$\langle\sigma_a\rangle=\frac{4\pi N(\operatorname{Im}\tau-|\tau|^2)}{k^2|1-(N-1)\tau Q|^2} \quad (34)$$

which indeed vanishes, as the optical theorem demands, when the spherules are non-absorbing so that μ and hence η are real.

Within this approximation, multiple scattering can be neglected provided

$$(N-1)|\tau Q|\ll 1. \quad (35)$$

Use of (3) gives

$$Q=\frac{1}{k^2R^2}\int_0^\infty dx\, x^2 p(x)\frac{\exp(ikRx)}{x}\frac{\sin kRx}{x}. \quad (36)$$

With the distribution (6) and (7) this can be integrated exactly and with (1) and $N\gg 1$ leads to

$$N\tau Q=\frac{2i\tau[(D+1)D]^{D/2}}{(2ka)^D(2-D)(D-1)}\left\{\left(\frac{b^2}{4k^2R^2}+1\right)^{1-D/2}\right.$$
$$\left.\times\exp\left[-i(2-D)\tan^{-1}\left(\frac{2kR}{b}\right)\right]-\left(\frac{b}{2kR}\right)^{2-D}\right\}. \quad (37)$$

When $kR\ll 1$, (19) and (20) give

$$|N\tau Q|\approx\frac{2^{1-D}|\mu^2-1|[(D+1)D]^{D/2}}{3(D-1)}(ka)^{3-D}\left(\frac{2kR}{b}\right)^{D-1}\ll 1 \quad (38)$$

confirming what is obvious, that multiple scattering is unimportant for small clusters of small spherules. The results for large clusters are less obvious; when $kR\gg 1$, (37) gives

$$\left.\begin{aligned}|N\tau Q|&\approx\frac{2^{1-D}[(D+1)D]^{D/2}|\mu^2-1|(ka)^{3-D}}{3(2-D)(D-1)}, &&\text{if } D<2\\ |N\tau Q|&\approx\frac{D(D+1)|\mu^2-1|kaN^{1-2/D}}{6(D-2)(D-1)}, &&\text{if } D>2\ .\end{aligned}\right\} \quad (39)$$

The limit for $D<2$ is independent of N and predicts that, provided ka is small enough, multiple scattering is negligible at all stages of aggregation. The limit for $D>2$ shows that multiple scattering becomes important when $N \sim (ka)^{-D/(D-2)}$, i.e. $kR \sim (ka)^{-(3-D)/(D-2)}$. When $D=3$ this is $(\mu^2-1)kR \sim 1$ which is the correct condition for the scattering from a homogeneous refracting ball to depend on more than the single scattering from its separate parts. The dependences of $|N\tau Q|$ on ka and N in (39) survive for any $p(x)$ satisfying (4) and (5) (and decaying faster than x^{-1} as $x \rightarrow \infty$); that is, they do not depend on the particular form (6).

Intuitive arguments support the distinction between $D<2$ and $D>2$. When $D<2$ the cluster is geometrically transparent (projected area is $N\pi a^2$ rather than πR^2); when $D>2$ it is geometrically opaque. And if we pretend to assign a bulk 'refractive index' μ_c to the cluster, according to the elementary rule

$$\mu_c \sim 1 + 4\pi\left(\frac{N}{R^3}\right)\frac{\tau}{k^3} \sim 1 + \frac{4\pi(\mu^2-1)}{N^{3/D-1}} \tag{40}$$

then the phase accumulated by a 'ray' passing through the cluster is

$$kR(\mu_c - 1) \sim \tfrac{4}{3}\pi|\mu^2-1|kaN^{1-2/D}, \tag{41}$$

which vanishes as $N \rightarrow \infty$ if $D<2$ and grows if $D>2$ (but note that the 'focal length' $R/(\mu_c-1)$ diverges even if $D>2$: we do not expect fractal balls to form images for any $D<3$. Finally we observe that $D=2$ is the dimension of a Brownian trail [12] and in the quite different problem of scattering by fractal surfaces [13] those whose profiles were graphs of Brownian motion also separated different sorts of diffraction.

Nevertheless we cannot rule out the possibility that when $D<2$ the fluctuations in the matrix elements of (26), which our mean-field approximation has neglected, might conspire to cause non-negligible multiple scattering when N exceeds some large negative D-dependent power of ka.

5. Scalar waves: single scattering formulae

For single scattering from the N-cluster the amplitude is (25) with $c_j=1$. Straightforward averaging over the directions of separations between spherules gives, for the differential cross-section for scattering by θ:

$$\left\langle \frac{d\sigma}{d\Omega}(\theta)\right\rangle = N\frac{d\sigma^{(1)}}{d\Omega}\left(1+(N-1)\int_0^\infty dx\, x^2 p(x)\frac{\sin(2kRx\sin(\theta/2))}{2kRx\sin(\theta/2)}\right) \tag{42}$$

where $d\sigma^{(1)}/d\Omega = |\tau|^2/k^2$ is the (isotropic) differential cross-section from a single spherule. With the fractal $p(x)$ given by (6) and (7) the cluster average can be evaluated exactly

$$\left\langle \frac{d\sigma}{d\Omega}(\theta)\right\rangle \Big/ N\frac{d\sigma^{(1)}}{d\Omega} \equiv F_d(\theta) = 1 + \frac{(N-1)\sin\{(D-1)\tan^{-1}[2kR/b\sin(\theta/2)]\}}{(D-1)(1+4k^2R^2\sin^2(\theta/2)/b^2)^{(D-1)/2}2kR\sin(\theta/2)/b}. \tag{43}$$

This is the factor by which the scattering per spherule is enhanced by coherence due to the fractality. In the forward direction there is complete coherence, as for any structure, and in fact

$$F_d(\theta) \approx N \quad \text{when} \quad \theta \ll (kR)^{-1}. \tag{44}$$

The large-angle scattering, however, saturates with coagulation (increasing N) to the value

$$F_d(\theta) \approx \frac{\sin[(D-1)\pi/2]}{(D-1)(2ka\sin(\theta/2)/b)^D} \quad \text{when} \quad \theta \gg (kR)^{-1}. \tag{45}$$

This saturation arises from the fractal structure and would not happen for homogeneously disordered spherules: when $D=3$ the limit (45) gives zero, whereas in reality (43) shows that the large-angle scattering decreases as $N^{-1/3}$. The main properties of $F_d(\theta)$ are illustrated in figure 1. The asymptotic angle dependence (45) has been verified experimentally, using light and X-ray scattering, for systems with a variety of fractal dimensions [14, 15].

The scattering cross-section is the integral of (43) over directions, and the enhancement factor F_s for the scattering per spherule is easily calculated (from (43) or from (33) as $1+(N-1)\,\mathrm{Im}\,Q$ with Q given by 37)

$$\langle\sigma_s\rangle/N\sigma_s^{(1)} \equiv F_s = 1 + \frac{2(N-1)}{(D-1)(2-D)X^2}\{(1+X^2)^{1-D/2}\cos\{(2-D)\tan^{-1}X\}-1\} \tag{46}$$

where $X \equiv 2kR/b$. For small clusters this has the trivial limit

$$F_s \approx N \quad \text{when} \quad kR \ll 1, \quad \text{i.e.} \quad N \ll (ka)^{-D}. \tag{47}$$

For large clusters the limiting behaviour depends on D:

$$\left.\begin{aligned} F_s &\approx \frac{2\cos[(2-D)\pi/2]}{(D-1)(2-D)(2ka/b)^D}, && \text{if} \quad D<2; \\ F_s &= \frac{\ln(4k^2a^2N/b)}{(2ka/b)^2}, && \text{if} \quad D=2; \\ F_s &\approx \frac{2N^{1-2/D}}{(D-1)(2-D)(2ka/b)^2}, && \text{if} \quad D>2; \end{aligned}\right\} \tag{48}$$

when $kR \gg 1$, i.e. $N \gg (ka)^{-D}$.

The full curves in figure 2 illustrate the N- and D-dependence of F_s. Just as for the multiple scattering estimates of § 4, what matters is whether $D<2$ or $D>2$. If $D<2$ the scattering per particle is enhanced by a factor which saturates to a large ($\sim(ka)^{-D}$) value that is independent of N. If $D>2$, coagulation continues to increase the enhancement factor, until the single-scattering approximation breaks down when $N \sim (ka)^{-D/(D-2)}$ (cf. (39)). $D=2$ shows the logarithmic behaviour typical of crossover dimensionalities.

The absorption cross-section is, from (34) (with $Q=0$) and (21), simply the sum of the absorptions of all the spherules, namely

$$\langle\sigma_a\rangle = N\sigma_a^{(1)} = (4\pi N/k^2)\,(\mathrm{Im}\,\tau - |\tau|^2) N\pi a^2 \tfrac{4}{3}ka\,\mathrm{Im}\,\mu^2. \tag{49}$$

If the real and imaginary refractive indices are of comparable magnitudes, absorption greatly exceeds scattering, even for large clusters, by a factor which from (48), (46) and (21) is of order $1/(ka)^{3-D}$ if $D<2$ and $1/(N^{1-2/D}ka)$ if $D>2$ (which is less than unity only when N exceeds the single-scattering limit after (39)). For smoke the extinction by fractal clusters of non-transparent spherules is therefore dominated by absorption.

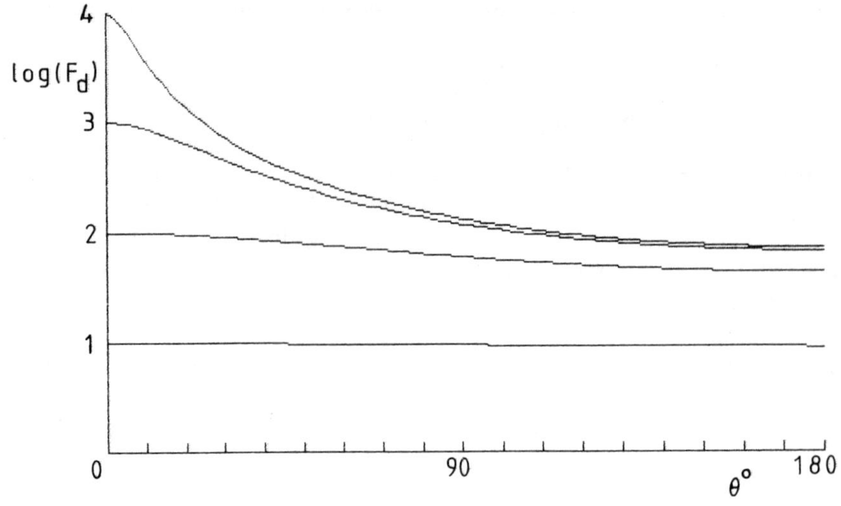

Figure 1. Coherent enhancement factor $F_d(\theta)$ for the differential cross-section per spherule (equation (43)), for $D=1{\cdot}78$, $ka=0{\cdot}111$ and $N=10$, 100, 1000, 10000, showing isotropic scattering for small N (cluster smaller than wavelength), and concentrated forward peak and saturation of large-angle scattering for large N. Logarithms are to base 10.

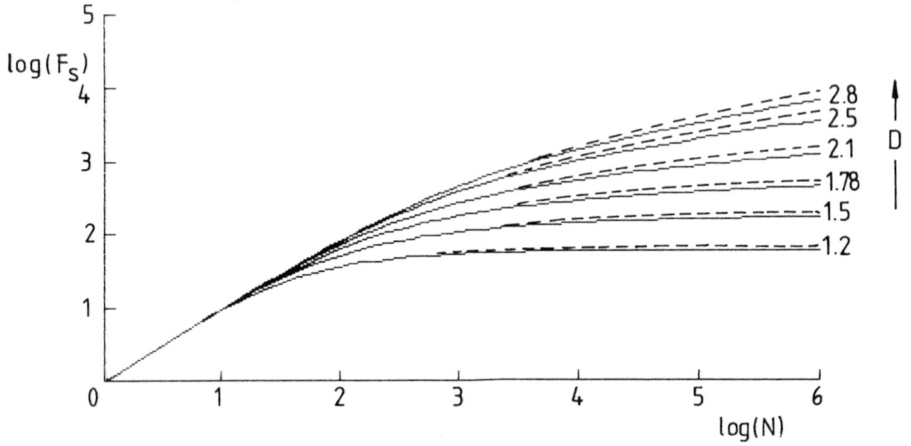

Figure 2. Coherent enhancement factors F_s for the scattering cross-section per spherule for scalar waves (full curves, from (46) and (1) and vector waves (broken curves, from (74) and (1)), for $ka=0{\cdot}111$, showing saturation with increasing N when $D<2$, and continuing increase when $D>2$. Logarithms are to base 10.

It is worth comparing the scattering and absorption for large ($kR \gg 1$) fractal clusters with $D<2$ (for example, smoke) with the corresponding predictions that would follow from the false assumption that coagulation produces large absorbing solid spheres ($D=3$). On this assumption the extinction cross-section of an N-cluster is given by geometrical optics as

$$\sigma_e^{\text{large solid}} = \pi R_{\text{solid}}^2 = \pi a^2 N^{2/3}. \tag{50}$$

The fractal extinction cross-section, given by (49), exceeds this by a factor

$$\langle\sigma_{\mathrm{a}}\rangle/\sigma_{\mathrm{e}}^{\text{large solid}}=\tfrac{4}{3}\,\mathrm{Im}\,\mu^2N^{1/3}ka=\tfrac{4}{3}\,\mathrm{Im}\,\mu^2kR_{\text{solid}}. \tag{51}$$

This exceeds unity for large solid spheres, so that the solid-coagulation assumption underestimates the extinction. A similar estimate for scattering shows that this is overestimated, except for *very* large clusters ($N>(ka)^{-3(4-D)}$) when fractals scatter more.

6. Vector waves: scattering equations

In this and the next two sections we follow the scalar-wave theory as closely as possible. The electronmagnetic fields satisfy not (12) but Maxwell's equations, and we will work entirely with the electric field $\mathbf{E}(\mathbf{r})$. Instead of (13) the incident field, with polarization unit vector $\mathbf{e}_{\mathrm{i}}(\perp\mathbf{k}_{\mathrm{i}})$, is

$$\mathbf{E}_{\mathrm{i}}=\mathbf{e}_{\mathrm{i}}\exp(\mathrm{i}\mathbf{k}_{\mathrm{i}}\cdot\mathbf{r}) \tag{52}$$

The cluster produces from this a scattered wave whose asymptotic form, involving a vector scattering amplitude, is

$$\mathbf{E}_{\mathrm{s}}=\frac{\exp(\mathrm{i}kr)}{r}\mathbf{f}(\mathbf{k}_{\mathrm{s}},\mathbf{k}_{\mathrm{i}}). \tag{52 a}$$

The scattering and absorption cross-sections are given by (16) and (18), the differential and extinction cross-sections being [16]

$$\frac{\mathrm{d}\sigma}{\mathrm{d}\Omega}(\mathbf{k}_{\mathrm{s}},\mathbf{k}_{\mathrm{i}})=\mathbf{f}(\mathbf{k}_{\mathrm{s}},\mathbf{k}_{\mathrm{i}})\cdot\mathbf{f}^*(\mathbf{k}_{\mathrm{s}},\mathbf{k}_{\mathrm{i}}) \tag{53}$$

and

$$\sigma_{\mathrm{e}}=(4\pi/k)\,\mathrm{Im}\,\mathbf{f}(\mathbf{k}_{\mathrm{i}},\mathbf{k}_{\mathrm{i}})\cdot\mathbf{e}_{\mathrm{i}}. \tag{54}$$

The fundamental approximation $ka\ll 1$ implies that each spherule scatters weak dipole radiation. At this stage we do not use the Rayleigh (i.e. Born) approximation because this fails to satisfy the optical theorem for real index μ. Instead we use a unitarity-preserving weak-scattering approximation to the exact Mie theory [17], in which a single spherule at the origin produces the scattered wave

$$\mathbf{E}_{\mathrm{s}}=\frac{3\rho\exp(\mathrm{i}kr)}{2kr}\left[(\mathbf{n}\wedge\mathbf{e}_{\mathrm{i}})\wedge\mathbf{n}+[3\mathbf{n}(\mathbf{n}\cdot\mathbf{e}_{\mathrm{i}})-\mathbf{e}_{\mathrm{i}}]\left(\frac{1}{(kr)^2}-\frac{\mathrm{i}}{kr}\right)\right] \tag{55}$$

where $\mathbf{n}\equiv\mathbf{r}/r$ and

$$\rho\equiv\sin\delta\exp(\mathrm{i}\delta) \tag{56}$$

in which δ is a p-wave phase shift given approximately, by solving Maxwell's equations, as

$$\delta\approx\tfrac{2}{3}(ka)^3\left(\frac{\mu^2-1}{\mu^2+2}\right). \tag{57}$$

For the single-spherule scattering amplitude the first (radiative) term in (55) gives

$$f_1(\mathbf{k}_{\mathrm{s}},\mathbf{k}_{\mathrm{i}})=\frac{3\rho}{2k^3}(\mathbf{k}_{\mathrm{s}}\wedge\mathbf{e}_{\mathrm{i}})\wedge\mathbf{k}_{\mathrm{s}}=\frac{3\rho}{2k}\left(\mathbf{e}_{\mathrm{i}}-\frac{(\mathbf{k}_{\mathrm{s}}\cdot\mathbf{e}_{\mathrm{i}})\mathbf{k}_{\mathrm{s}}}{k^2}\right). \tag{58}$$

Instead of (21), the three cross-sections for complex μ are found using (16) and (54) to be

$$\frac{d\sigma^{(1)}}{d\Omega}(\mathbf{k}_i, \mathbf{k}_s) \approx k^4 a^6 \left|\frac{\mu^2-1}{\mu^2+2}\right|^2 [1-(\mathbf{k}_s \cdot \mathbf{e}_i)^2/k^2],$$

$$\sigma_s^{(1)} \approx \pi a^2 \tfrac{8}{3}(ka)^4 \left|\frac{\mu^2-1}{\mu^2+2}\right|^2, \quad \sigma_a^{(1)} \approx \sigma_e^{(1)} \approx \pi a^2 4ka \,\mathrm{Im}\left(\frac{\mu^2-1}{\mu^2+2}\right). \tag{59}$$

For unpolarized incident light, the directional factor in $d\sigma^{(1)}/d\Omega$ becomes $\frac{1}{2}(1+\cos\theta)$.

For the cluster the asymptotic scattered wave, analogous to (22), is, from (55),

$$\mathbf{E}_s = \frac{3\rho}{2k^3 r} \sum_{j=1}^{N} (\mathbf{k}_s \wedge \mathbf{E}_j) \wedge \mathbf{k}_s \exp(-i\mathbf{k}_s \cdot \mathbf{r}_j) \tag{60}$$

where $\mathbf{E}_j$ is the total field incident on the jth spherule. This in turn is the sum of the incident field and those scattered to j from the other spherules:

$$\mathbf{E}_j = \mathbf{e}_i \exp(i\mathbf{k}_i \cdot \mathbf{r}_j) + \frac{3\rho}{2} \sum_{l \neq j} \left[(\mathbf{n}_{jl} \wedge \mathbf{E}_l) \wedge \mathbf{n}_{jl} + [3\mathbf{n}_{jl}(\mathbf{n}_{jl} \cdot \mathbf{E}_l) - \mathbf{E}_l] \times \left(\frac{1}{(kr_{jl})^2} - \frac{i}{kr_{jl}} \right) \right] \frac{\exp(ikr_{jl})}{kr_{jl}}. \tag{61}$$

Now we introduce multiple scattering coefficients $\mathbf{d}_j$ (cf. (24)) by

$$\mathbf{E}_j \equiv \mathbf{d}_j \exp(i\mathbf{k}_i \cdot \mathbf{r}_j). \tag{62}$$

This gives finally the vector-wave equations, exact for small ka,

$$\mathbf{f}(\mathbf{k}_s, \mathbf{k}_i) = \frac{3\rho}{2k^3} \sum_{j=1}^{N} (\mathbf{k}_s \wedge \mathbf{d}_j) \wedge \mathbf{k}_s \exp[i(\mathbf{k}_i - \mathbf{k}_s) \cdot \mathbf{r}_j] \tag{63}$$

with $\mathbf{d}_j$ determined by

$$\mathbf{d}_j = \mathbf{e}_i + \frac{3\rho}{2} \sum_{l \neq j} \left[(\mathbf{n}_{jl} \wedge \mathbf{d}_l) \wedge \mathbf{n}_{jl} + [3\mathbf{n}_{jl}(\mathbf{n}_{jl} \cdot \mathbf{d}_l) - \mathbf{d}_l] \left(\frac{1}{(kr_{jl})^2} - \frac{i}{kr_{jl}} \right) \right] \exp(ikr_{jl} - i\mathbf{k}_i \cdot \mathbf{r}_{jl}). \tag{64}$$

7. Vector waves: multiple scattering estimates

By analogy with the scalar case, we estimate the effects of multiple scattering by making the mean-field approximation that all multiple scattering coefficients are the same, and parallel to the incident polarization, i.e.

$$\mathbf{d}_j \approx d\mathbf{e}_j. \tag{65}$$

This amounts to replacing by their average values the matrix elements in (64), whose component along $\mathbf{e}_i$ gives (cf. (30) and (31))

$$d = [1 - \tfrac{3}{2}(N-1)\rho P]^{-1} \tag{66}$$

where

$$P \equiv \left\langle \frac{\exp(ikr_{jl})}{kr_{jl}} \cos \mathbf{k}_i \cdot \mathbf{r}_{jl} \left[1 - (\mathbf{n} \cdot \mathbf{e}_i)^2 + [3(\mathbf{n} \cdot \mathbf{e}_i)^2 - 1] \left(\frac{1}{(kr_{jl})^2} - \frac{i}{(kr_{jl})^2} \right) \right] \right\rangle \tag{67}$$

with $\mathbf{n} \equiv \mathbf{r}_{jl}/r_{il}$.

Again this approximation preserves the optical theorem exactly. This is more surprising than in the scalar case, because P in (67) involves the near-zone (non-radiative) terms in the dipole field, and the scattering involves only the far field. To show it, we first write the extinction cross-section using (63) as

$$\langle\sigma_e\rangle=\frac{6\pi N}{k^2}\operatorname{Im}\rho d=\frac{6\pi N[\operatorname{Im}\rho+\frac{3}{2}|\rho|^2(N-1)\operatorname{Im}P]}{k^2|1-\frac{3}{2}(N-1)\rho P|^2}. \tag{68}$$

The scattering cross-section is

$$\langle\sigma_s\rangle=\frac{9|\rho|^2|d|^2}{4k^2}\sum_{j=1}^{N}\sum_{l=1}^{N}\left\langle\iint d\Omega_s\exp[i(\mathbf{k}_i-\mathbf{k}_s)\cdot\mathbf{r}_{lj}]\left[1-\left(\frac{\mathbf{k}_s\cdot\mathbf{e}_i}{k}\right)^2\right]\right\rangle$$

$$=\frac{6\pi N|\rho|^2\left(1+[3(N-1)/8\pi]\left\langle\iint d\Omega_s\exp(-i\mathbf{k}_s\cdot\mathbf{r}_{jl})\{1-[\mathbf{k}_s\cdot\mathbf{e}_i/k]^2\}\cos(\mathbf{k}_i\cdot\mathbf{r}_{jl})\right\rangle\right)}{k^2|1-\frac{3}{2}(N-1)\rho P|^2}. \tag{69}$$

An elementary calculation shows that the angular integral is precisely $4\pi\operatorname{Im}P$, so the absorption cross-section is

$$\langle\sigma_a\rangle=\frac{6\pi N(\operatorname{Im}\rho-|\rho|^2)}{k^2|1-\frac{3}{2}(N-1)\rho P|^2} \tag{70}$$

which indeed vanishes for non-absorbing spherules.

Within this approximation, multiple scattering can be neglected provided

$$\tfrac{3}{2}(N-1)|\rho P|\ll 1. \tag{71}$$

Use of (9) gives

$$P=\frac{1}{k^2R^2}\int_0^\infty dx\,x^2p(x)\frac{\exp(iX)}{x^2}\left\{\left[1-\frac{2}{X^2}+\frac{3}{X^4}+\frac{i}{X}\left(1-\frac{3}{X^2}\right)\right]\right.$$
$$\left.\times\sin X+\frac{1}{X}\left(1-\frac{3}{X^2}+\frac{3i}{X}\right)\cos X\right\} \tag{72}$$

where $X\equiv kRx$. Note that the isotropy of the cluster statistics has eliminated the dependence on the initial polarization $\mathbf{e}_i$. P is similar to the scalar-wave quantity Q (equation (36)) with the factor $\{\ \}$ replacing $\sin X$. Now, as $X\to 0$ the factor $\{\ \}$ vanishes like X, just as $\sin X$ does. Therefore the integral in (72) converges for fractal cluster distributions $p(x)$, and this implies that the vector-wave condition (71) depends on N, D and ka in the same way as the scalar-wave condition (35), apart from numerical factors (in the next section we shall see that these numerical factors are closely similar). The discussion based on (37)–(39) can therefore be taken over to the vector case, the main conclusion being that as for scalar waves multiple scattering can always be neglected when $D<2$ (provided of course that $ka\ll 1$).

9. Vector waves: single scattering formulae

For single scattering from the N-cluster the amplitude is (63) with $\mathbf{d}_j=\mathbf{e}_i$. Straightforward cluster averaging shows that the differential cross-section is given exactly by the scalar-wave formula (42) (the single-spherule differential cross-section being (59)), and the same discussion applies, based on (43)–(45) and figure 1.

The scattering cross-section is (69) with the modulus factor unity. The angular integral is $2\pi\,\mathrm{Im}\,P$ and (72) and the distribution (6) and (7) lead to

$$\langle\sigma_s\rangle/N\sigma_s^{(1)}\equiv F_s \tag{73}$$

where the enhancement factor F_s for the scattering per spherule is now

$$F_s=1+\frac{3(N-1)}{(D-1)X^2}[\![(1+X^2)^{1-D/2}\{\cos[(2-D)\tan^{-1}X]\}(T_0+T_2/X^2+T_4/X^4) \\ +\sin[(2-D)\tan^{-1}X](T_1/X+T_3/X^3)-(U_0+U_2/X^2+U_4/X^4)]\!] \tag{74}$$

in which $X\equiv 2kR/b$ and

$$\left.\begin{aligned}
&T_0\equiv\frac{1}{2-D}-\frac{2}{4-D}+\frac{2}{6-D}, && T_1\equiv\frac{4}{3-D}-\frac{4}{4-D}-\frac{8}{5-D}+\frac{8}{6-D},\\
&T_2\equiv\frac{2}{2-D}-\frac{4}{3-D}-\frac{10}{4-D}+\frac{24}{5-D}-\frac{12}{6-D}, && T_3\equiv\frac{8}{3-D}-\frac{24}{4-D}+\frac{24}{5-D}-\frac{8}{6-D},\\
&T_4\equiv\frac{2}{2-D}-\frac{8}{3-D}+\frac{12}{4-D}-\frac{8}{5-D}+\frac{2}{6-D}, && u_0\equiv\frac{1}{2-D},\\
&U_2\equiv\frac{2}{2-D}-\frac{4}{3-D}+\frac{2}{4-D}, && U_4\equiv\frac{2}{2-D}-\frac{8}{3-D}+\frac{12}{4-D}-\frac{8}{5-D}+\frac{2}{6-D}.
\end{aligned}\right\} \tag{75}$$

This is the vector equivalent of (46), and it has similar limiting behaviour. For small clusters, much cancellation leads to

$$F_s\approx N \quad \text{when} \quad kR\ll 1, \quad \text{i.e.} \quad N\ll(ka)^{-D}. \tag{76}$$

For large clusters, the limits depend on D:

$$\left.\begin{aligned}
&F_s\approx\frac{3\cos[(2-D)\pi/2]}{(D-1)(2-D)(2ka/b)^D}\left(1-\frac{4(2-D)}{(4-D)(6-D)}\right), && \text{if} \quad D<2,\\
&F_s\approx\frac{3\ln(4k^2a^2N/b)}{2(2ka/b)^2}, && \text{if} \quad D=2,\\
&F_s\approx\frac{3N^{1-2/D}}{(D-1)(D-2)(2ka/b)^2}, && \text{if} \quad D>2,
\end{aligned}\right\} \tag{77}$$

when $kR\gg 1$, i.e. $N\gg(ka)^{-D}$.

The dashed curves in figure 2 illustrate the N- and D-dependence of F_s. Evidently the vector and scalar scattering cross-sections are quantitatively as well as qualitatively similar, and the same remarks (following equation (48)) apply. The absorption cross-section is, likewise, simply the sum of the absorptions of all the spherules, namely

$$\langle\sigma_a\rangle=N\sigma_a^{(1)}=\frac{6\pi N}{k^2}(\mathrm{Im}\,\rho-|\rho|^2)=N\pi a^2 4ka\,\mathrm{Im}\left(\frac{\mu^2-1}{\mu^2+2}\right). \tag{78}$$

This is similar to (49) and the discussion from that equation to the end of § 5 applies here too.

10. Conclusions

Our theory shows that for both scalar and vector waves the scattering cross-section of a D-dimensional fractal cluster of N small spherules is different for $D>2$ and $D<2$. For $D>2$ the scattering per spherule for large clusters grows with aggregation as $N^{1-2/D}$. For $D<2$ (smoke) the surprising result is that the scattering per spherule increases to a constant saturation value which exceeds that of a solitary spherule by a factor of order $(ka)^{-(3-D)}$. The absorption cross-section is independent of D and simply the sum of absorptions of the separate spherules.

Any real smoke is an assembly of clusters of different sizes, with N as a random variable. If the smoke has n spherules per unit volume and has coagulated into a sparse distribution of n/N clusters with, on the average, N spherules each, then the absorption or scattering length (for exponential attenuation) of an incident wave, is

$$L=[(n/N)\sigma]^{-1} \tag{79}$$

where σ is the appropriate cross-section. Our results show that for absorption L is always independent of N, that is, of coagulation, and for scattering from smoke ($D<2$) L decreases with N to a constant value, unaffected by further coagulation. These predictions could be tested by experiments in which the absorption and scattering were measured for smoke coagulating under controlled conditions.

In obtaining these results, multiple scattering has been neglected. This was justified (for all N when $D<2$ and for $N\lesssim(ka)^{-D/(D-2)}$ when $D>2$) by estimating the multiple scattering using a mean field theory. Obviously this is an approximation and it would be very interesting to have an exact theory of the large N limit for fractal clusters, based (probably) on a renormalization analysis of the field fluctuations within a cluster.

Acknowledgments

We would like to thank D. D. Drysdale, G. W. Mulholland, D. J. Rabash, S. Simons and M. Williams for helpful communications. This research was not supported by any military agency.

References

[1] Chylek, P., Ramaswamy, V., Cheng, R., and Pinnick, R. G., 1981, *Appl. Optics*, **20,** 2980–2985.
[2] Forrest, S. R., and Witten, T. A., 1982, *J. Phys.* A, **12,** L109–L117.
[3] Mountain, R. D., and Mulholland, G. W., 1984, in *Kinetics of Aggregation and Gelation: Proceedings of the International Topical Conference*, edited by F. Family and D. P. Landau (Amsterdam: North-Holland).
[4] Crutzen, P. J., and Birks, J. W., 1982, *Ambio*, **11,** 114–125.
[5] Turco, R. P., Toon, O. B., Ackerman, T. P., Pollack, J. B., and Sagan, C., 1983, *Science*, **222,** 1283–1292.
[6] Pittock, A. B., Ackerman, T. P., Crutzen, P. J., MacCracken, M. C., Shapiro, C. S., and Turco, R. P., 1985, in *Environmental Consequences of Nuclear War* (SCOPE 28), Vol. 1, *Physical and Atmospheric Effects* (Chichester, New York: J. Wiley).
[7] Greene, O., Percival, I., and Ridge, I., 1985, *Nuclear Winter* (Cambridge: Polity Press).
[8] Messiah, A., 1962, *Quantum Mechanics* (Amsterdam: North-Holland).
[9] Jones, A. R., 1979, *Proc. R. Soc.* A, **366,** 111–127.
[10] Ziman, J. M., 1979, *Models of Disorder* (Cambridge University Press).
[11] Warner, M., and Gubernatis, J. E., 1984, *Phys. Rev.* B, **32,** 6347–6357.

[12] MANDELBROT, B. B., 1982, *The Fractal Geometry of Nature* (San Francisco: Freeman).
[13] BERRY, M. V., 1979, *J. Phys.* A, **12,** 781–797.
[14] SCHAEFER, D. W., MARTIN, J. E., WILTZIUS, P., and CANNELL, D. S., 1984, *Phys. Rev. Lett.*, **52,** 2371–2374.
[15] WEITZ, D. A., HYANG, J. S., LIN, M. Y., and SUNG, J., 1985, *Phys. Rev. Lett.*, **54,** 1416–1419.
[16] JACKSON, J. D., 1975, *Classical Electrodynamics* (New York: Wiley).
[17] VAN DE HULST, H. C., 1957, *Light Scattering by Small Particles* (New York: Wiley).

Reprinted with permission from *Chemical Physics Letters* 107, J.S. Newhouse, P. Argyrakis, and R. Kopelman, "Fractal to Classical Crossover of Chemical Reactions," pp. 48–52, ©1984 Elsevier Science Publishers B.V., North-Holland Publishing.

FRACTAL TO CLASSICAL CROSSOVER OF CHEMICAL REACTIONS *

J.S. NEWHOUSE, P. ARGYRAKIS ‡ and R. KOPELMAN
Department of Chemistry, The University of Michigan, Ann Arbor, Michigan 48109, USA

Received 17 February 1984

Computer simulations on binary reactions of random walkers (A + A → A) on two- and three-dimensional percolation clusters bear out the recent superuniversality conjecture (integrated reaction rate ∝ $t^{2/3}$). Moreover, the fractal-to-euclidean crossover ($t^{2/3}$ to t dependence) parallels that of the single walker.

1. Introduction

There has been much recent interest in the fractal aspects of heterogeneous chemical reactions [1–6]. While most work considered the trapping rate of excitons or electrons at impurity or defect centers [1–4], i.e. where only one reactant moves (the "trap" is fixed) there has also been some specific work on binary chemical reactions where both reactants are free to move, i.e. genuine bimolecular reactions [5–7]. For the simplest such reaction,

$$A + A \rightarrow \text{products}, \tag{1}$$

it has been shown via simulation [6] (confirming an earlier conjecture [5]) that the reaction of eq. (1), when proceeding on a fractal matrix, obeys the following integrated rate equation:

$$\rho^{-1}(t) - \rho^{-1}(0) = K_0 t^f, \tag{2}$$

where $\rho(t)$ is the reactant density at time t, K_0 a constant and

$$f = \tfrac{1}{2} d_s \tag{3}$$

where d_s is the *spectral dimension* ("fracton" dimension) [8]. For a euclidean lattice (homogeneous medium), $d_s = 2$, i.e. $f = 1$, and eq. (2) reduces to the classical (textbook) result:

$$\rho^{-1}(t) - \rho^{-1}(0) = K_0 t. \tag{4}$$

Eq. (4) tells us that the *rarefication of the reactant species A* (described by the left-hand side) *proceeds linearly in time.* However, for disordered media of fractal nature, eq. (2) tells us that this rarefication proceeds with a fractal power of the time, prescribed by the spectral dimension of the fractal.

Experimental results for exciton fusion (triplet–triplet annihilation: T + T → products) on binary lattices [5,6,9] have shown a behavior obeying eq. (2), and *not* eq. (4), under appropriate conditions (in refs. [5,9] the differential rate equation leading to eq. (2) has been used). However, microscopically disordered media, such as random binary lattices, have an exact fractal nature [8,10] only under very restricted conditions, i.e. at the critical percolation concentration C_c. The question then arises: How widely spread and important is the fractal regime for chemical reactions. The present work supplies some quantitative answers and shows that the regime of fractal behavior is not limited to some exotically restricted conditions.

The logical premise here (as in ref. [5]) has been that the reaction in a system of identical random walkers (A) will reflect the motion of a single random walker: The same power f that determines the exploration range $S(t)$ of a single random walker,

$$S(t) \propto t^f, \tag{5}$$

will also determine the reaction kinetics (e.g. eq. (2)). We note that, on a discrete lattice or matrix, $S(t)$ re-

* Supported by NIH Grant No. 2 R01 NS80116-16 and NATO Grant No. SA 52505 RG 295/83.
‡ Permanent address: Department of Physics, University of Crete, Iraklion, Crete, Greece.

0 009-2614/84/$ 03.00 © Elsevier Science Publishers B.V. (North-Holland Physics Publishing Division)

Volume 107, number 1 CHEMICAL PHYSICS LETTERS 18 May 1984

duces to $S(N)$, the mean number of distinct sites visited in N steps:

$$S(N) \propto N^f . \qquad (5')$$

The conjecture that f is given by the spectral dimension d_s (eq. (3)) [8,10] has been confirmed by many simulations [11]. For percolation lattices this conjecture only holds [8–11] at the percolation threshold (C_c). It has, however, been conjectured that there exists a *crossover* from fractal to euclidean behavior that is related to the percolation correlation length (ξ) [4,12]. This crossover can be crudely expressed [11] by letting the exponent f vary from *2/3* (the magic value at percolation thresholds in all dimensions [8]) to *unity* (the value for three-dimensional lattices and continua). Here we give quantitative simulation results for the *single random walker crossovers.* We then show that the *crossovers of reacting systems of random walkers* are completely analogous (i.e. the dependence of f on the concentration and on the time).

2. Simulations and results

The methods of the Monte Carlo simulations have been described before [6,11,13,14]. Figs. 1 and 2 show the *single-walker exploration* versus time for square (two-dimensional) and simple cubic (three-dimensional) lattices, respectively. In each case the low-

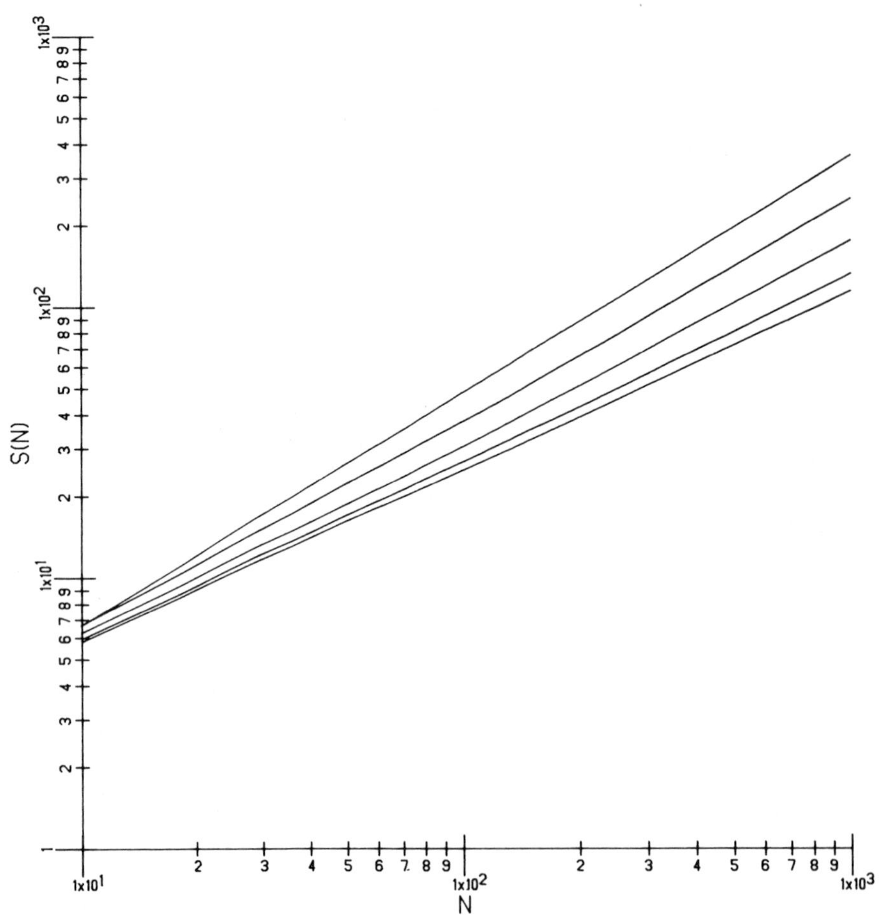

Fig. 1. S_N versus N for a two-dimensional square lattice. Top to bottom: C = 1.00, 0.80, 0.70, 0.65, and 0.60. Averages of 1000 runs, except for the 0.60 case where 2290 runs were retained out of 3283 runs, in order to discard the walks on finite clusters [11, 13].

Volume 107, number 1 CHEMICAL PHYSICS LETTERS 18 May 1984

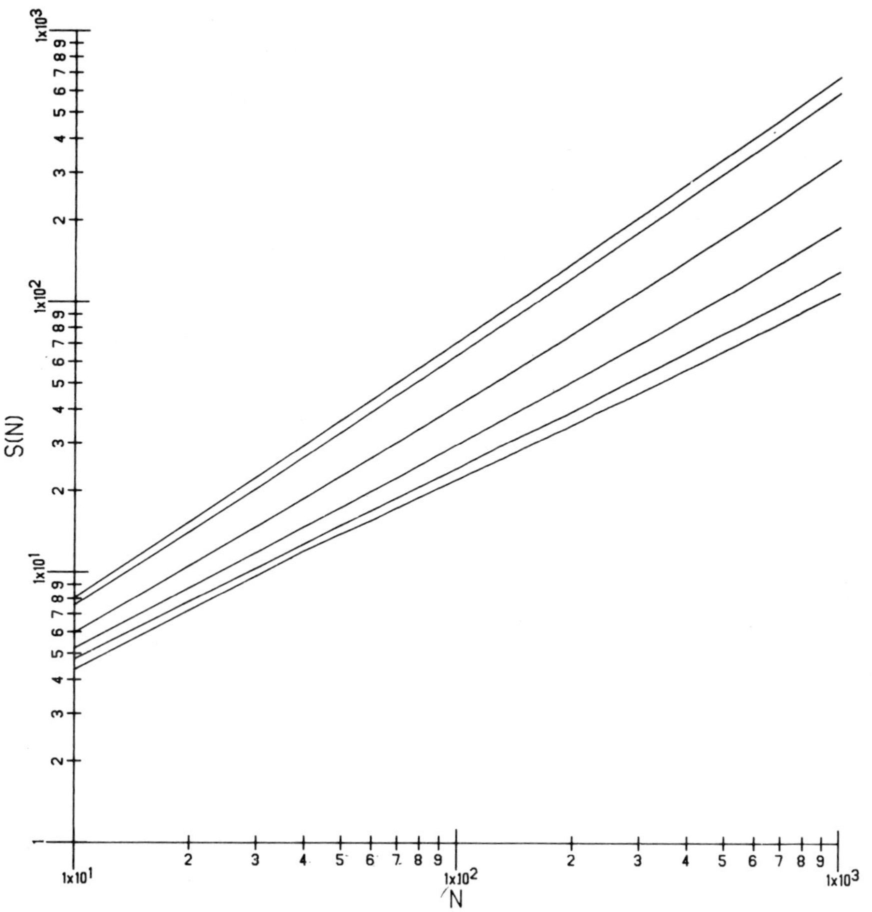

Fig. 2. S_N versus N for a three-dimensional simple cubic lattice. Top to bottom: C = 1.00, 0.80, 0.50, 0.40, 0.36 and 0.32. Averages of 1000 runs except for the 0.32 case where 501 runs were retained out of 1000, in order to discard the walks on finite clusters [11,13].

est curve is close to the threshold (critical concentration C_c) and the higher curves are for higher concentrations, up to unity (pure crystal). On both figures the lowest curves are quite straight and have a slope close to 2/3. The highest curve for the three-dimensional case (C = 1.0) is also fairly straight, with a slope close to unity (only the three-dimensional case should have a slope of strictly unity, and this only at $t \to \infty$, because of the logarithmic correction involved in the two-dimensional random walk [15]). We would particularly like to emphasize that the lines for the intermediate concentrations are curved, not straight. For this regime, at early times, the slope is lower, closer to the fractal value (2/3) while at longer times it is higher, close to the euclidean value (1).

Note that all "crossover" lines curve in the "upward" direction. A quantitative crossover description, in terms of effective f values, is given in table 1.

Figs. 3 and 4 show the *reacting walker rarefication* versus time, again for square (two-dimensional) and simple cubic (three-dimensional) lattices, respectively. The behavior is very similar to that mentioned for the single walker. The lowest curves (at about C_c), show straight lines with slopes very close to 2/3 (see table 1 for exact values), while the highest curves show, again, fairly straight lines with slopes close to unity (especially for the three-dimensional case). Again, the curves for the intermediate concentrations show a "crossover" from lower (about 2/3) to higher (below 1) values, and again these lines curve in the

Volume 107, number 1 CHEMICAL PHYSICS LETTERS 18 May 1984

Table 1
Effective f values

C	Single walker		Reacting walkers	
	two-dim. (slopes from fig. 1)		two-dim. (slopes from fig. 3)	
	overall f	asymptotic f	overall f	asymptotic f
1.00	0.889	0.883	0.920 [a]	0.935 [a]
0.80	0.839	0.822	0.865	0.866
0.70	0.768	0.785	0.806	0.831
0.65	0.713	0.722	0.748	0.780
0.60	0.665	0.671	0.676	0.674
	three-dim. (slopes from fig. 2)		three-dim. (slopes from fig. 4)	
	overall f	asymptotic f	overall f	asymptotic f
1.00	0.988	0.989	0.981	0.980
0.80	0.981	0.987	0.976	0.967
0.70			0.970	0.980
0.50	0.935	0.945	0.941	0.961
0.40	0.850	0.863	0.861	0.912
0.36	0.757	0.786	0.748 [b]	0.763 [b]
0.32	0.709	0.677	0.674	0.687

[a] Ref. [16]. [b] $C = 0.35$.

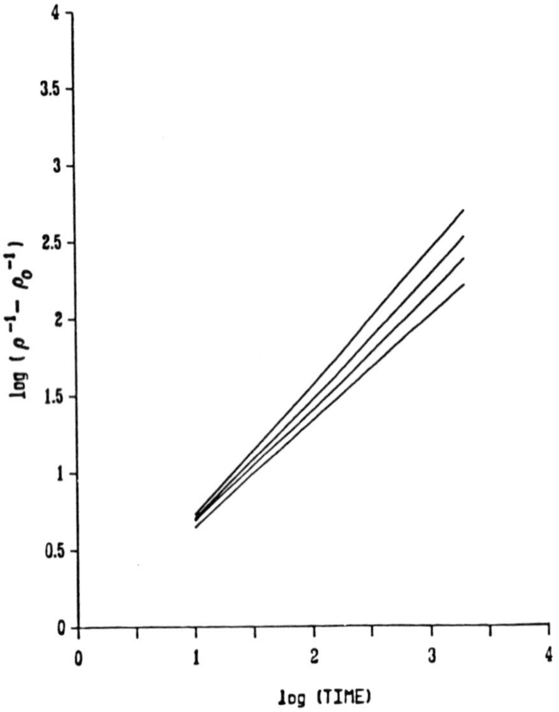

Fig. 3. $\text{Log}[\langle\rho^{-1}(t)\rangle - \langle\rho^{-1}(0)\rangle]$ versus $\log t$ for a two-dimensional square lattice. Top to bottom: $C = 0.80$, 0.70, 0.65, 0.60. Averages of 100 runs, 2000 steps each, on 200×200 lattices, with cyclic boundary conditions, $\rho(0) = 0.02$. Forced random walks, limited to largest cluster only.

"upward" direction. A quantitative crossover description is given in table 1, columns 4 and 5. This compares very favorably with table 1, colums 2 and 3, for the single walks.

3. Summary

Evidently there is a very satisfactory analogy between the single and the reacting multiple random walk systems. This bears out earlier conjectures [5,6] and demonstrates the usefulness of the concept of *fractal rate constant and fractal-to-classical crossover.* Most striking, perhaps, is the fact that for the system of reacting random walkers, we find $f = 0.67$ for the two-dimensional case and 0.69 for the three-dimen-

Volume 107, number 1 CHEMICAL PHYSICS LETTERS 18 May 1984

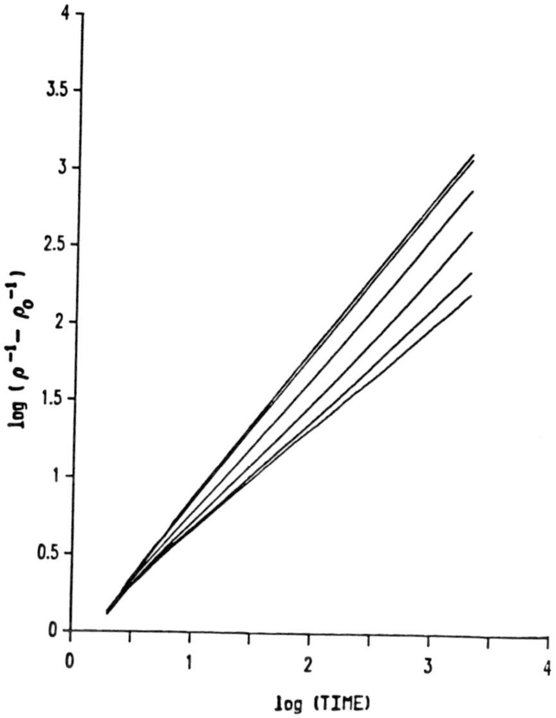

Fig. 4. Log $[\langle\rho^{-1}(t)\rangle - \langle\rho^{-1}(0)\rangle]$ versus log t for a three-dimensional simple cubic lattice. Top to bottom: $C = 1.00$, 0.80, 0.50, 0.40, 0.35, 0.32. Averages of 1000 runs, 2000 steps each, on 40 × 40 × 40 lattices, with cyclic boundary conditions, $\rho(0) = 0.02$. Forced random walks, limited to largest clusters only.

sionl case, in excellent agreement with the single random walker result of $f = 0.67$ and 0.68, respectively, and the superuniversality conjecture of 2/3.

References

[1] P.G. de Gennes, Compt. Rend. Acad. Sci. (Paris) A296 (1983) 881.
[2] P. Evesque, J. Phys. (Paris) 44 (1983) 1217.
[3] J. Klafter and A. Blumen, J. Chem. Phys. 80 (1984) 875; A. Blumen, J. Klafter and G. Zumofen, Phys. Rev. B28 (1983) 6112; B27 (1983) 3429; J. Klafter, A. Blumen and G. Zumofen, J. Phys. Chem. 87 (1983) 191.
[4] I. Webman, Phys. Rev. Letters 52 (1984) 220.
[5] P. Klymko and R. Kopelman, J. Phys. Chem. 87 (1983) 4565.
[6] R. Kopelman, P.W. Klymko, J.S. Newhouse and L. Anacker, Phys. Rev. B29 (1984) 3747.
[7] P. Meakin and H.E. Stanley, J. Phys. A17 (1984) L173.
[8] S. Alexander and R. Orbach, J. Phys. (Paris) 43 (1982) L625.
[9] P.W. Klymko, Ph.D. Thesis, University of Michigan (1984).
[10] S. Alexander, Ann. Israel Phys. Soc. 5 (1983) 149.
[11] P. Argyrakis and R. Kopelman, Phys. Rev. B29 (1984) 511, and references therein.
[12] Y. Gefen, A. Aharony and S. Alexander, Phys. Rev. Letters 50 (1983) 77.
[13] P. Argyrakis, K. Anacker and R. Kopelman, J. Stat. Phys. (1984), to be published.
[14] L. Anacker, J.S. Newhouse and R. Kopelman, J. Stat. Phys. (1984), to be published.
[15] E.W. Montroll and G.W. Weiss, J. Math. Phys. 6 (1965) 167.
[16] L. Anacker, private communication.

Reprinted with permission from *Percolation Structures and Processes,* edited by G. Deutscher, R. Zallen and J. Adler, "Linear Problems on Percolation Clusters," S. Alexander, pp. 149–166, ©1983 Annals of the Israel Physical Society.

CHAPTER 7

LINEAR PROBLEMS ON PERCOLATION CLUSTERS

S. Alexander

The Racah Institute of Physics
The Hebrew University, Jerusalem*

and

Institute for Theoretical Physics
Univ. of Calif., Santa Barbara, CA 93106

* Permanent address

Annals of the Israel Physical Society, v. 5
© Israel Physical Society 1983

Percolation Structures and Processes
Eds. G. Deutscher, R. Zallen and J. Adler

ABSTRACT

The mapping of linear equations described by a Laplacian on percolation clusters is discussed. Using scaling considerations we show that the mapping is described by the fracton dimensionality which seems to have the universal value of 4/3 for percolation clusters. We review scaling results for the diffusion problem and for the density of states and describe the vibrational density of states for a system for which percolation describes the quenched disorder. Explicit mapping of the equations is treated using a network model for the clusters and their behaviour under dilation is discussed. Some of the implications for localisation are also discussed and it is shown that the fractal geometry induces a new type of localisation which does not require disorder.

1. INTRODUCTION

The purpose of this article is to discuss the solution of standard equations on percolation clusters. The canonical problem of this type is deGennes' ant[1] - i.e. the properties of a random walk confined to percolation clusters. It is however fairly straight-forward to see that there is a large class of related and closely analogous problems. One can consider solutions of the Schrodinger equation (for free particles or in a tight binding model), the mechanical properties of such clusters, superconducting properties (e.g. solutions of the Landau Ginzburg equations)[2-5] and various electric circuit realizations. The interest in this problem stems largely from the fact that percolation is, by far, the best understood type of quenched disorder. It is directly relevant to some physical situations and can be applied to others by a suitable self consistent definition of the percolation bonds[6-8]. The usual approach to most of the problems listed is somewhat different. Thus, for example, the electronic properties are usually discussed in the framework of models with a random potential and no explicit geometrical disorder. We only have the geometrical disorder induced by the structure of the percolation clusters. This introduces some novel features but is, in other respects, equivalent.

There are a number of simplifiying features to the percolation problem.

First one is dealing with a purely geometrical type of disorder. The percolation problem defines a geometrical structure (a network) and one is looking for solutions of standard equations on these networks.

The second point is that percolation is a critical phenomenon. Thus standard scaling and renormalisation group ideas apply. A geometrical interpretation of these relationships implies an anomalous fractal geometry at short distances[1], a cluster size distribution[10] and a crossover to a homogeneous geometry at the connectivity coherence length[9-11] of the percolation problem.

We shall only discuss problems (such as those listed above) which are described by Laplacians or by second order finite difference operators. Thus most of our discussion will be concerned with the implications of mapping such operators on the anomalous fractal geometry of the percolation clusters. We shall discuss the implications for the density of states[12] and briefly for the structure of the eigenfunctions[4,19].

Two types of approaches have been used. First one can use a direct scaling approach. In essence this amounts to treating scale transformations as a symmetry operation. On fractal geometries this has non-trivial implications because one requires more than one dimensionality to characterize such geometries. The mapping of a Laplacian on a self similar fractal is described by an intrinsic fractron dimensionality (D_f)[12], relevant to eigenstate counting and e.g. the density of states, which is usually smaller than the Haussdorf dimension (D). For percolation clusters D_f seems to have a universal value (4/3) independent of the dimensionality of space at least for $d \geq 2$ [12,13].

For many purposes one would like to have an explicit algorithm for introducing the anomalous geometries into the equations. A technique for doing this is to regard the percolation clusters as networks of one dimensional strands connected at junctions or ending at (dead) end points. Percolation defined on a lattice automatically has this form. On a network all linear second order equations can be reduced to linear finite difference equations for the junction amplitudes[4] which can then be handled explicitly (or by decimation).

In section 2 we discuss diffusion on percolation clusters using a scaling approach. We shall follow the approach of Gefen e. a.[11] based on older ideas of Straley[14]. We shall then calculate the density of states for a single fractal cluster[12].

In section 3 we use crossover considerations to determine the relevant indices for the percolation problem. Following Rammal and Toulouse[13] we discuss the geometrical implications of the empirical fact that the fracton dimensionality retains its mean field value. In section 4 we discuss the total density of states for a percolation system[15].

The explicit scaling argument could in principle be repeated for other problems. Instead we demonstrate the equivalence in section 5 by mapping the different equations on networks[4]. It can then be seen that the equations all have the same structure and therefore behave in the same way under scale transformations.

We derive the equations and discuss their behaviour under scale transformations. We also discuss some of the implications for the structure of the eigenfunctions.

The list of references we give is incomplete and probably involves a considerable personal bias. Most of the results quoted are fairly recent and came to my attention as preprints or orally. I apologize if I have lost track of some of these results or failed to trace their publications.

2. DIFFUSION ON THE CRITICAL INFINITE CLUSTER

We consider a random walk on a self similar fractal - e.g. the infinite cluster at the critical percolation concentration p_c. Since there is no geometrical length scale the only length in the problem is the (time dependent) diffusion length $\lambda(t)$. Thus the probability of finding a particle which started at the origin at t=0 at position $\underline{r}$ at time t should have a scaling form:

$$P(\underline{r},t) \propto \lambda^{-D} f(r/\lambda) \tag{2.1}$$

where f(x) is a function which depends only on the geometry of the fractal and D is the Haussdorf dimension. Eqn. 2.1 does of course hold for a Cartesian continuum. In that case ($\lambda \propto t^{1/2}$), f(x) is a Gaussian and D=d the Cartesian dimension. In general we only assume the scaling form. Following Straley[14] we first consider the scaling properties of the second moment of P(r,t). We write [11,12]:

$$\langle r^2 \rangle_t \propto t^{2/(2+\theta)}; \quad \lambda(t) \propto t^{1/(2+\theta)} \tag{2.2}$$

This can be regarded as a definition of the diffusion index θ.

It is instructive to consider the meaning of θ from a different point of view. Equation 2.2 is equivalent to a size dependent effective diffusion constant

$$D_{eff}(r) \propto r^{-\theta} \tag{2.3}$$

A straightforward dimensional argument now relates θ to the scaling behaviour of the conductance[4,11]. Let[9]

$$\sum (r) \propto r^{\mu} \tag{2.4}$$

be the conductance of a (fractal) piece of size r. Then

$$\theta = 2 + \mu - D \tag{2.4}$$

An explicit derivation of this relationship in terms of the master equation on a network [4] is given in section 4. We note that for a Cartesian space μ=d-2; D=d) one finds $\theta = 0$ as required. We shall also see below (in section 3) that crossover arguments for the percolation problem give the same result.

The relationship of θ to the scaling behaviour of the conductance emphasizes the fact that it is a new index, essentially independent of the Haussdorf dimensionality (D). One can construct many different fractals all having the same D but quite different conductances (i.e. μ).

Combining equations 2.1 and 2.2 one finds the probability that random walker finds himself at the origin at time t[12]

$$P(o,t) \propto \lambda^{-D} \propto t^{-D/2+\theta} \tag{2.6}$$

This is a very important quantity. Its immediate interpretation is of course that it measures the weight of loops starting and ending at the origin among all random walks[16]. One also notes that P(r,t) is the Green's function for the diffusion problem. P(o,t) is therefore the diagonal element of this function. It is related to the density of states through a Laplace transform[17,18]:

$$\underline{P}(o,t) = \int_{o}^{\infty} N(\varepsilon) e^{-\varepsilon t} d\varepsilon \tag{2.7}$$

where $N(\varepsilon)$ should be understood as a suitably averaged, density of states for the eigenstates of a master equation or a diffusion equation on the percolation cluster. Following reference 12 we define a fracton dimension

$$D_f = 2D/(2+\theta) \tag{2.8}$$

In terms of D_f one can write

$$P(o,t) \propto t^{-D_f/2} \tag{2.6a}$$

so that the fracton dimensionality (D_f) (rather than D) replaces the Cartesian dimension (d) in this expression. The same holds for the density of states. Combining 2.6a with 2.7 one finds[12]

$$N(\varepsilon) \propto \varepsilon^{((D_f/2)-1)} \tag{2.9}$$

This has the standard form except for the fact that the relevant dimension is the fracton dimension.

A different way of stating this is to say that D_f is the dimension relevant to the mapping of a gaussian fixed point on the fractal geometry (of the percolation cluster). This is the natural formulation in following the behaviour of the equations under dilation of the self similar geometry[4] (see e.g. the analysis of the Sierpinski gasket in ref. 19).

As shown by Mandelbrot[20] a fractal is invariably characterised by more than one dimensionality. The answers to different questions which are all described by the Cartesian dimension, require different and, in principle, independent dimensions for a fractal. Like the topological dimension D_f is an intrinsic geometric property of the fractal and does not depend on the way it is embedded in Cartesian space. Thus the fracton dimensionality for fractal lines is equal to the topological dimension (i.e. $D_f \equiv 1$) independent of D and d[12].

3. FRACTON DIMENSION FOR PERCOLATION CLUSTERS

Both D and θ can be determined from simple scaling and crossover arguements on the infinite cluster above the percolation threshold ($p > p_c$). The essential assumption is that the short distance geometry ($r < \xi_p$) of the infinite cluster is fractal and described by the same indices as at p_c. This is equivalent to standard scaling assumptions for critical phenomena. Consider first the Haussdorf dimensionality.

We follow references 9 and 10 and compare the estimates of the mass of the infinite cluster in a volume of size ξ_p from the small scale (fractal) and long distance (homogeneous) limit

$$\xi_p^{\ D} \simeq (p-p_c)^\beta \cdot \xi_p^d \tag{3.1}$$

Using $\xi_p \propto (p-p_c)^{-\nu}$ this gives[9,10]

$$D = d - \beta/\nu \tag{3.2}$$

An analogous argument for the crossover of the diffusion constant[11] uses equation 2.3:

$$D_{ic}(p) \simeq D(\xi_p) \propto \xi_p^{-\theta} \tag{3.3}$$

where D_{ic} is the diffusion constant for a particle <u>known</u> to be on the infinite cluster[1]

$$D_{ic} \propto (p-p_c)^{t-\beta} \tag{3.4}$$

Thus

$$\theta = (t - \beta)/\nu \tag{3.5}$$

where t is the conductivity index. Both D and θ vary considerably between $d = 2$ and the upper critical dimensionality ($d = 6$). It is therefore quite surprising that the fracton dimensionality seems to be independent of d. One finds[12]

$$D_f = 4/3; \quad 2 \lesssim d \tag{3.6}$$

within the accuracy of the available exponents. Recent more accurate results for the conductivity index t[21, 22] suggest that the numerical accuracy of eqn. 3.6 is considerably higher than suggested by the values (of ref. 10) originally used in reference 12. We therefore believe that it is probably exact. One notes that this amounts to a new, additional, scaling relation which determines θ (or equivalently the conductivity index t) in terms of the critical indices of the percolation problem[12]

$$t = \frac{1}{2}((3d-4)\nu - \beta) \qquad (3.7)$$

We note that our crossover arguments assume hyperscaling (i.e. a single percolation length scale). Thus the argument breaks down in the mean field regime ($d > 6$) where one expects the mean field values ($D = \theta = 4$) but the relationships 3.2 and 3.7 are no longer valid.

A very interesting geometric interpretation of equation 3.6 has been presented by Rammal and Toulouse[24,25]. They consider the number of distinct sites visited by a random walk of N steps (S(N)). These sites necessarily form a connected set on the underlying geometry. One can then define the boundary of this set. There are B(N) sites in the set with accessible neighbours outside the set. Thus

$$\frac{dS}{dN} \propto B/S \qquad (3.8)$$

This gives the correct result for the relationship between S(N) and N in cartesian space ($B \propto S$; $S(N) \propto N$, $d > 2$; $B = C$; $S(N) \propto N^{1/2}$, $d = 1$). Now S(N) is obviously bounded by N and by the diffusion volume

$$S(N) < N; \quad N^{D_f/2} \qquad (3.9)$$

For $D_f < 2$ each site (in the diffusion volume) is visited many times ($N \gg N^{D_f/2}$) so that one can assume

$$S(N) \simeq N^{D_f/2} \qquad (3.10)$$

Rammal and Toulouse[24] now note that $D_f = 4/3$ is equivalent to

$$B(N) \propto [S(N)]^{1/2} \propto N^{D_f/4} \qquad (3.11)$$

Substitution in equation 3.8 leads to

$$S^{3/2} \propto N \qquad S \propto N^{2/3} = N^{D_f/2} \qquad (3.12)$$

This is a very curious result and requires some comment. Explicitly it means that the number of sites on the diffusion front is proportional to $S^{1/2}$. Naively one would tend to determine the "front" from the maximum in the derivative of

$P(r,t)$ (i.e. $\partial f/\partial x = 0$ - eqn. 2.1). This defines a sphere with a radius proportional to $\lambda(N)$. The implication of 3.11 would be that sphere of radius R, which contains R^D sites onthe critical infinite cluster, cuts the cluster at $B(R) \propto R^{D/2}$ points. This is however geometrically impossible[26]. As demonstrated by Mandelbrot[20] such a sphere cuts the fractal cluster in R^{D-1} points with probability one. If one substitutes thisin the Rammal Toulouse equation 3.8 one finds, using 3.10 and the definition of D_f (eqn. 2.8)

$$D_f = 2D/(D + 1); \quad \theta = D - 1 \tag{3.13}$$

in clear contradiction with experiment.

The origin of this discrepancy is in the twofold averaging process implied in the definition of $P(r,t)$ (eqn. 2.1). From obvious symmetry considerations $P(r,t)$ can depend only on the magnitude r. However for any specific choice of origin ($\underline{r}_o$) the conditional probability density $P(\underline{r}_o,o;\ \underline{r}_o + \underline{r},t)$ is not spherically symmetric (in $\underline{r}$) but rather reflects the specific geometrical correlations of the cluster associated with this choice of origin. The diffusion rate is that of the individual realisations andthe average of the $B(\underline{r}_o;\ N)$ cannot be determined from the average of the $P(\underline{r}_o,o;\ \underline{r}_o + \underline{r},\ t)$ $(P(r,t))$. This difficulty is fairly universal in disordered systems (see e.g. ref. 7). For the problem at hand it arises even for uniform recursive fractals. Thus equation 3.13 is not even valid for fractal lines or for the Sierpinski gasket[12].

It would be very instructive to see some numerical check of the convoluted diffusion front implied by equation 3.11. One notes that D/2 can be both larger and smaller than D-1. We also note that one expects related problems with other averages where they would show up in a failure of the law of large numbers. One could for example look at the average $\langle r^2\rangle_{\underline{r}_o,\Omega}$ in some specified direction. The average over $\underline{r}_o$ is well defined

$$\langle\langle r^2\rangle_{\underline{r}_o,\Omega}\rangle_{\underline{r}_o} = \langle r^2\rangle \propto \lambda^2(N)$$

but fluctuations remain large so that the variance

$$V = \langle(\langle r^2{}_{\underline{r}_o,\Omega} - \langle r^2\rangle)^2\rangle_{\underline{r}_o} / \langle r^2\rangle^2$$

remains finite. One also expects $f(r/\lambda)$ to decay much more slowly around $r/\lambda = 1$ than in the isotropic case.

4. THE DENSITY OF STATES

It is interesting to consider the total density of states of a system described by percolation as a quenched disorder. This requires taking proper account of the cuttoffs introduced into the spectrum by the finite cluster size distribution[10] and crossovers (at ξ_p) on the infinite cluster. As pointed out earlier we assume that the short range properties ($r < \xi_p$) are fractal and therefore are described by the indices D, θ and D_f. The long range behaviour is assumed isotropic. This neglects statistical fluctuations, and is certainly not an adequate description in the crossover region. Following Alexander, Laermans, Orbach and Rosenberg[15] we discuss the (mechanical) vibration spectrum. Since the equations involve a second derivative one has $\epsilon \to \omega^2$ (ω the vibrational frequency) and[12]

$$N(\omega) \propto \omega^{D_f-1} \tag{4.1}$$

replacing eqn. 2.9. Introducing a fracton Debye frequency (Ω_f) in the usual way one has

$$N_f(\omega) = D_f\, \omega^{(D_f-1)}/(\Omega_f)^{D_f} \tag{4.2}$$

normalised to the basic atomic unit.

Consider first a finite cluster of size $L(<\xi_p)$. The frequency spectrum has a lower cutoff when the characteristic length of the modes λ (analogous to the diffusion length) is of order L. This gives

$$\omega_{min}(L) \propto L^{-(2+\theta)/2} \tag{4.3}$$

Integrating 4.2 from ω_{min} to Ω_f one gets

$$\int_{\omega_{min}}^{\Omega_F} N_f(\omega)d\omega = 1 - \left(\frac{\omega_{min}}{\Omega_f}\right)^{D_f} \tag{4.4}$$

Using eqn. 4.3 the correction term becomes

$$(\omega_{min}/\Omega_f)^{D_f} \propto (a/L)^D \tag{4.5}$$

where a is a microscopic length. Thus the whole cluster (of volume $\sim L^D$) misses one mode. Physically this is associated with the rigid motion of the cluster as a whole.

On the infinite cluster we have a crossover at ξ_p

$$\omega_{co} \propto \xi_p^{-(2+\theta)/2} \tag{4.6}$$

Below ω_{co} one has normal phonons and

$$N_{ph}(\omega) = d\frac{\omega^{d-1}}{\Omega_D d}; \quad \omega < \Omega_{co} \tag{4.7}$$

per volume a^d.

We have defined an apparent Debye frequency

$$2\pi c/a = \Omega_D = (a/\xi_p)^{\theta/2} \cdot \Omega_f \tag{4.8}$$

Since Ω_f should be associated with the microscopic masses and force constants it must be comparable to the Debye frequency of the completely bonded system ($p \simeq 1$). Thus Ω_D and the velocity of sound (c) are greatly reduced if $\xi_{p/a}$ is large. Above ω_{co} the states on the infinite cluster are again described by eqn. 4.2.

Consider now the total density of states $N_t(\omega)$. We have:

$$N_t^p(\omega) = N_{ic}^p(\omega) = d \cdot \omega^{d-1}/\Omega_D^d; \quad \omega < \omega_{co} \tag{4.9a}$$

at low frequencies. This integrates to one mode per volume ξ_p^d. At higher frequencies one has a contribution from the infinite cluster and from all finite clusters large enough to maintain a mode of given frequency ($\omega > \omega_{min}(L)$). A straightforward integration using the Stauffer distribution[10,11] gives

$$N_t^f(\omega) = N_{ic}^f(\omega) + N_{fc}^f(\omega) = (D_f/\omega)(\omega/\Omega_f)^{(2d/2+\theta)}; \quad \omega > \omega_{c.o.} \tag{4.9b}$$

The result of the integration over the distribution of clusters is to replace D_f by $\frac{2d}{2+\theta}$ ($> D_f$) in eqn. 4.9b. We still miss the center of mass modes which have no frequencies within the harmonic model. They constitute a fraction

$$P_{c.o.m.} = (d-D)/d = \beta/\nu d \tag{4.10}$$

of all modes and have the mass distribution of the finite clusters.

Thus one predicts a crossover in the vibrational density of states at ω_{co}. The density of states has different power laws on the two sides of the crossover. There is also a discontinuity in the density of states at ω_{co}:

$$N^p(\omega_{co})/N^f(\omega_{co}) = d/D_f > 1 \tag{4.11}$$

In addition one has the center of mass modes which would certainly give rise to unharmonic "two level" modes if one somehow went beyond the harmonic percolation mode.

5. NETWORK MODELS

Consider bond percolation defined on a lattice. We can consider the bonds as one dimensional connections between the lattice point. This defines a network consisting of 1 dimensional strands ending at junction points where at least three strands meet or at a free dangling end. We shall call both types of end points junctions below.

We can map a continuum problem on this net in two ways.

a) By writing tight binding equations or a master equation on the net. For example the master equation:

$$\frac{dp_i}{dt} = \sum_{\text{bonds}} (p_j - p_i) \tag{5.1}$$

where the summation on the right hand side is only over the open bonds of the percolation problem.

b) One can consider the strands as a one dimensional continuum and solve the continuum equations on them with suitable boundary conditions at the junctions.

Procedure b was suggested by deGennes[3] for the Landau Ginzburg equations. For linear second order equations we have shown[4] that it is convenient to eliminate the strands and set up finite difference equations involving only the relevant amplitudes at the junctions. An explicit procedure for case b is discussed in reference 4 and can easily be generalized to case a.

To be specific consider the Schrodinger equation. On the strands we have the 1-d equation

$$q^2\psi(s) = -(\partial^2\partial s^2)\psi(s) \tag{5.2}$$

where s is a running coordinate along the strand and we have written q^2 for the eigenvalues. The amplitudes have to be continuous at the junctions and, in addition, one has the condition on the outgoing derivatives along the strands:[3,4]

$$\sum_j (\partial/\partial s_{ij})\, \psi_{ij}(s) = 0 \tag{5.3}$$

where the summation is over all strands whose ends are connected at i. Solving eqn. 5.2 in terms of the values at the ends (ψ_i, ψ_j) and substituting in eqns. 5.3 one ends up with the difference equations:[4]

$$q^2 m_i \psi_i + \sum_j \sigma_{ij}(\psi_j - \psi_i) = 0 \tag{5.4}$$

where

$$m_i = q^{-1} \sum_j \tan(q\ell_{ij}/2); \quad \sigma_{ij} = q[\sin(q\ell_{ij})]^{-1} \tag{5.5}$$

and ℓ_{ij} is the length of the strand connecting i and j. If we had started with case a we would have obtained equations of the same form with a different definition of the constants namely:

$$\sigma_{ij} = \sin q/\sin(qN_{ij}) \tag{5.6}$$

and

$$m_i = [\, \sum \tan(qN_{ij}/2) - (Z_i-2)\tan(q/2)] \cdot \frac{\sin q}{q^2} \tag{5.7a}$$

for equations of the form 5.1. One obtains a slightly different form for m_i for the electronic tight binding problem:

$$m_i = [\, \sum \tan(qN_{ij}/2) - (Z_i - 2)\cot q] \cdot \frac{\sin q}{q^2} \tag{5.7b}$$

In equations 5.6 and 5.7 Z_i is the number of nearest neighbours bonded to i and N_{ij} the length of the strand connecting i and j.

Explicit solutions of eqns. 5.4 on simple networks, lattices and the Sierpinski gasket have been discussed in the literature[4,19]. They could certainly also be solved numically on simulations of percolation clusters but, to our knowledge, no such solution has been attempted. For our purposes the main interest is in the behaviour of the equations under dilation. One chooses a suitable subset of the vertices and eliminates all the other ψ_i from the equations. For a self-similar fractal the new equations should have the same structure as the original ones. An explicit example of a recursive fractal (the Sierpinski gasket) is worked out in refs. 4 and 19. We shall follow reference 4 in applying scaling considerations to the equations on percolation clusters.

Consider first the scaling behaviour of the m_i. For lattice vibrations they play the role of a mass and therefore scale with the Haussdorf dimension (for small q). One notes that short dead ends, which are eliminated by the dilation, will contribute to the m_i (but not to the σ_{ij}). Thus a dilution which increases the separation of vertices by a factor b will also increase m

$$m \rightarrow b^D m \tag{5.8}$$

Similarily, for q = 0, one obtains (from 5.4) the equations for a resistor network. The σ_{ij} scale like the bond conductances. If one continues the dilation on the infinite cluster up to ξ_p, the σ_{ij} become the Skal-Shklovskii-deGennes[27] bond conductances. Thus

$$\sigma \rightarrow b^{(D - \theta -2)} \cdot \sigma \tag{5.9}$$

Finally if we interpret eqn. 5.4 as a master equation one finds for the diffusion constant the scaling behaviour

$$D \propto b^2\sigma/m \propto b^{-\theta} \tag{5.10}$$

confirming our results in sections 2 and 3.

The language in the above argument is somewhat vague. Obviously any microscopic model (e.g. eqns. 5.5-5.7) will not give one value of m and σ but rather some stactical distribution of the m_i and σ_{ij}. Thus the dilation process involves statistical averaging. For a random fractal one does not even expect that they converge to a single value. We are implicitly assuming that the distributions converge to some stationary shape, determined by the transformation, whose averages than scale as indicated. This implies that the details of the initial microscopic model are not important.

We would like to conclude with some comments on the structure of the eigenfunctions. The main point is that the fractal geometry in itself implies a strong tendency for localisation even in the absence of any random fluctuations. As a result we expect the length

$$\lambda(q) \propto q^{\frac{-1}{2+\theta}} \tag{5.11}$$

to play the role of a localisation length. While we have no proper proof for the validity of this claim we would like to present some heuristic arguments which we find convincing. The relationship of this type of localisation to that found using other models is not obvious. In essence one is presumably dealing with the same universal phenomenon but we believe that there are probably important model dependent differences in many details.

The most general argument is to apply the "gang of four" scaling relations[29] to a fractal. The argument is due to Toulouse[28]. From equation 5.9 we can calculate the β function of reference 29:

$$\beta(\sigma) = \frac{d(\ell n\sigma)}{d(\ell nL)} = D(1 - 2/D_f) \tag{5.12}$$

where we have used the definition of D_f (eqn. 2.8). Thus the sign of β is determined by the fracton dimension and will be negative whenever $D_f < 2$. In a way this is not surprising.

The argument of Abrahams e.a.[29] simply implies that any amount of disorder will be sufficient to induce localisation for $D_f = 4/3$ (i.e. on percolation clusters). This certainly implies that eigenfunctions on the critical infinite cluster at p_c are localized. Scaling then suggests that the localisation length must be proportional to $\lambda(q)$. If wave functions for different eigenvalues (q^2) are similar and differ only in the length scale (λ) the localisation length must also be proportional to λ.

From a different point of view one could look explicitly at the way the solutions of equations 5.4 behave under scale transformations. If the eigenfunctions are localized they should shrink and eventually disappear from the spectrum. The structure of the eigenfunctions has been studied in detail on the Sierpinski gasket[4,19]. We note that this is a recursive homogeneous fractal ($D = \ell n3/\ell n2 \simeq 1.6$ $D_f = \frac{2\ell n3}{\ell n5} \simeq 1.36$) with no disorder.

Domany e.a.[19] find that almost all eigenfunctions are localized and standard localisation criteria such as

$$\lim_{L \to \infty} \frac{\sum \psi_i^4}{(\Sigma\psi_i^2)^2} > 0 \tag{5.13}$$

are obeyed (L the size of the gasket). This is also reflected in the fact that only a very small fraction of the eigenfunctions are sensitive to the boundary conditions on a large gasket. For localized functions one also expects that they should shrink under scale transformations and eventually disappear when one reaches the lower cutoff. This is also found on the Sierpinski gasket. This certainly suggests that the fractal geometry gives rise to a strong tendency to localisation even in the absence of any disorder.

There are however some more delicate points which make it difficult to generalize the gasket results to percolation clusters which are only statistically self similar.

On the gasket one has exact self similarity so that the structure of the equations (5.4) is invariant under dilation. This gives rise to an exact iterative relationship between the eigenvalues. One also finds that the spectrum is bounded from below and from above and that the localized eigenstates are highly degenerate. On a fractal one only has self similarity in a statistical sense and this certainly makes a difference.

We only want to discuss one central point. On the gasket the scaling behaviour is associated with a fixed point of the scale transformation - essentially the mapping of a Gaussian fixed point. The whole spectrum is however controlled by an infinite set of points in the spectrum which contains all (finite and infinite) limit cycles of the transformation and those points which eventually are transformed into the limit cycle points (a Julia set[30]). This gives the spectrum chaotic features and also effects the range over which the scaling behaviour is relevant. It is not evident at present how these peculiarity would be reflected in a statistical fractal and whether they have important effects on the validity of the scaling results.

ACKNOWLEDGEMENT: I would like to thank G. Deutscher without whose encouragement insistence and patience this article would never have been written. I also thank R.Rammal,D.Stauffer,G. Toulouse and my collaborators in references 11,12,19,23 for many helpful discussions and private communications.

The hospitality of the Institute of Theoretical Physics and in particular of W. Kohn and P. Pincus who invited me there are gratefully acknowledged. This work was supported in part by the U.S. National Science Foundation contract number 4-444024-21944 and grant number PHY 77-27084.

REFERENCES

1. P.G. deGennes, La Recherche 7, 919 (1976).
2. P.G. deGennes, Comptes Rendues vol. 292, Series II, 9 (1981).
3. P.G. deGennes, Comptes Rendues, Vol. 292, series II, 279, 1981.
4. S. Alexander, Phys. Rev. B, 1983 (in press).
5. S. Alexander and E. Halevi, J. de Physique (1983).
6. V. Ambegaokar, B.I. Halperin and J.S. Langer, Phys. Rev. B. 4, 2612 (1971).
7. S. Alexander, J. Bernasconi, W.R. Schneider and R. Orbach, Rev. Mod. Phys. 53, 175 (1981).
8. S. Alexander, Phys. Rev. B 23 2951 (1981).
9. Y. Gefen, A. Aharony, B.B. Mandelbrot and S. Kir kpatrick, Phys. Rev. Lett. 47, 1771 (1981)
10. D. Stauffer, Phys. Repts. 54, 3 (1979).
11. Y. Gefen, A. Aharony and S. Alexander, Phys. Rev. Lett. 50 77 (1983)
12. S. Alexander and R. Orbach, J. de Physique (Lettres), 43 L625 (1982).
13. R. Rammal and G. Toulouse (preprint)
14. J.P. Straley, J. Phys. C 13, 2991 (1980)
15. S. Alexander, C. Laermans, R. Orbach and H.M. Rosenberg (1982 preprint).
16. S. Havlin, preprint.
17. S. Alexander, J. Bernasconi, W.R. Schneider and R. Orbach, J. de Physique C6, 706 (1978).
18. Section II of ref. 7.
19. E. Domany, S. Alexander, D. Ben Simon and L.P. Kadanoff, preprint; R. Rammal (private communication).
20. B.B. Mandelbrot, The Fractal Geometry of Nature, W.H. Freeman and Co., San Francisco (1982).
21. C.D. Mitescu (this volume) and note added in proof in reference 12.
22. D. Stauffer (private communication) for d = 3.
23. S. Alexander, T. Witten, G. Guest and H. Nakanishi (Preprint).
24. R. Rammal and G. Toulouse (preprint).
25. A closely related argument in terms of the diffusion equation itself has been developed by Y. Gefen (private communication).
26. I would like to thank A. Aharony for pointing this out to me.

Dynamics of Fractal Networks

R. ORBACH

Random structures often exhibit fractal geometry, defined in terms of the mass scaling exponent, *D*, the fractal dimension. The vibrational dynamics of fractal networks are expressed in terms of the exponent $\bar{\bar{d}}$, the fracton dimensionality. The eigenstates on a fractal network are spatially localized for $\bar{\bar{d}}$ less than or equal to 2. The implications of fractal geometry are discussed for thermal transport on fractal networks. The electron-fracton interaction is developed, with a brief outline given for the time dependence of the electronic relaxation on fractal networks. It is suggested that amorphous or glassy materials may exhibit fractal properties at short length scales or, equivalently, at high energies. The calculations of physical properties can be used to test the fractal character of the vibrational excitations in these materials.

THE CONCEPT OF FRACTAL STRUCTURE, DEVELOPED BY Benoit Mandelbrot (*1*), has potential utility because it is possible that structures that appear purely random can be described within a geometric mathematical framework. Fractal concepts may describe not only the static geometrical properties of such structures but also their dynamical properties and interactions with external measurement probes. The purpose of this article is to introduce the basic concepts behind the use of fractal geometry in performing these operations.

The fractal description of nature has developed so rapidly and so broadly in recent years, and fractal geometry has become such a powerful tool, that it would be impossible to attempt complete coverage of this subject. Therefore, only the more universal aspects of the structural properties of fractal networks will be analyzed in this article; more specific applications have been described (*2*).

The class of fractal structures treated in this article is limited to those that exhibit "self-similar" geometry. This means that the structure's geometrical properties are indistinguishable as a function of length scale (or resolution). Examining the fractal with ever more finely divided "rulers" would not result in any discernable difference in geometry. Conversely, the length scale of measurement could not be determined solely from observation of the fractal structure. This behavior usually breaks down at very short length scales, appropriate to the atomic or "building block" regime. In many physical systems (such as the percolating network, described later), the structure also ceases to be fractal at very long length scales, where it appears homogeneous or continuous. Mandelbrot (*1*) was the first to determine that many structures in nature exhibit self-similar geometry.

The most easily understood fractal concept is that of density. For fractal structures, this means that there is no constant relation between mass and volume as the length scale is changed. The amount of mass inside a sphere of radius r for a homogeneous (or Euclidean) structure scales as $M(r) = Ar^d$, where A is a numerical constant and d is the spatial or Euclidean dimension (d is 3 in our usual world). Analogously, the amount of mass inside a sphere of radius r for a fractal structure with self-similar geometry scales as $M(r) = Br^D$, where B varies according to the "lacunarity" (*1*) and D is the fractal dimension. In general, D is less than or equal to d because of the "open" character of fractal structures. That is, they tend to exhibit inhomogeneous arrangements of particles, with large amounts of open spaces (voids) and irregular atomic arrangements.

We are now in a position to calculate the density (ρ) of particles inside a sphere of radius r (that is, of volume $V(r) = Cr^d$, where C is a constant) for a fractal structure:

$$\rho(r) = M(r)/V(r) = Br^D/Cr^d \propto r^{D-d} \quad (1)$$

For self-similar structures, the fractal dimension D does not depend on r. This is an extraordinary result, for it implies an order in structures that to a casual observer appear to be completely disordered. It is even more remarkable because densities in seemingly unrelated phenomena in nature appear to behave in such a manner (*1*).

Because D is less than or equal to d, the density falls off with increasing length scale, implying that fractal objects of large size would be extraordinarily light. This must be tempered with the

R. Orbach is in the Department of Physics, University of California, Los Angeles 90024.

remark that at short length scales the atomic local geometries may not be fractal, while at very long length scales the geometry may appear to be homogeneous, implying an upper cutoff for r for those situations in which Eq. 1 may apply. Beyond this upper cutoff, $\rho(r)$ would simply be a constant.

An example where only the lower limit is appropriate can be found in aggregates of silica particles produced by flame hydrolysis of $SiCl_4$, known commercially as the products Cab-O-Sil and Alfasil (*3*) (here, the structure is Euclidean for short length scales and fractal for long length scales). From the results of electron microscopy studies, these materials are known to consist of ramified aggregates of very small, nearly spherical SiO_2 units ($R \sim 25$ Å) that stick together to form clusters measuring approximately 1000 Å across. Small angle neutron-scattering studies (*4*) on normal, compressed, and water-suspended powders, ranging in density from 0.009 to 0.45 g cm^{-3}, measure particle densities that obey Eq. 1 [the number density of SiO_2 units in a cluster diminishes with increasing size (r) of the volume sampled by the neutron-scattering study]. The fractal dimensionality D is found to be 2.52 ± 0.05 for all the samples studied. This is an explicit example where D is less than d ($d = 3$).

An example of a structure that has an upper limit for the length scale over which the structure is fractal is the much-studied percolating network (*5*) [here, the structure is fractal for short length scales (but greater than a fundamental grid dimension) and Euclidean for long length scales]. This model is thought to describe various physical structures, such as gels (*6*), polymers (*7*), and the transverse force constants of glassy materials (*8*). It forms the basis for studies of the flow of liquids through material aggregates [for example, oil and water through sandstone (*9*)]. There are two kinds of percolating networks: site and bond. To create a site-percolation network, each intersection of a d-dimensional grid is occupied at random with probability p. Sites are said to be connected if they are adjacent along a principal direction (that is, along a vertical or horizontal direction in $d = 2$ for a square grid, but not along the diagonals). A bond-percolation network is created by laying down the elementary components that make up the grid at random with probability p. The sites are then those points where the elementary components touch. The two kinds of percolation networks are not simply related, except for some special cases (*5*). We shall work with site percolation exclusively in this article.

Remarkably, a critical probability p_c exists such that, for $p \geq p_c$, a connected cluster will cross the grid continuously from one side to the other. This is referred to as an infinite cluster because, for an infinitely large grid, it will be of infinite extent. In contrast, the remainder of the occupied sites will form clusters of finite size. The probability of an occupied site being on the infinite network, $P(p)$, is given by the characteristic exponent β through the scaling formula

$$P(p) \propto (p - p_c)^{\beta} \quad (2)$$

We identify the lower cutoff for the percolating network to be a length a that forms the horizontal and vertical lengths that make up the elementary grid. An upper cutoff length exists for the percolating network (*5*); this is referred to as ξ_p. This quantity is usually termed the percolation correlation length, or the pair connectedness length for percolation. For $r \geq \xi_p$, the percolating network appears homogeneous, and the density ρ is a constant. The percolating network exhibits self-similar geometry for $a \leq r \leq \xi_p$. The occupied site density on the iinifinite cluster is given by Eq. 1 (the average occupied site density is simply p). We say, therefore, that the percolating network is fractal for length scales r between $a < r < \xi_p$ and Euclidean for $r \geq \xi_p$. The upper cutoff length ξ_p exhibits the scaling relation (*5*)

$$\xi_p \simeq a|p - p_c|^{-\nu_p} \quad (3)$$

For a site percolation network in $d = 2$, ν_p is 4/3 [this is the so-called den Nijs conjecture (*10*), now thought to be exact], β_p is 5/36 [also thought to be exact (*11*)], and p_c is 0.5928 [for bond percolation in $d = 2$, p_c is 0.5 (*5*)]. The fractal dimension D is approximately 1.9 (*12*).

For the site percolation networks ($d = 2$) pictured in Figs. 1 through 4 (*13*), the largest cluster is always shown in white, the second largest in bright yellow, the third in dull yellow, and successively smaller clusters in shades varying from orange to red to light blue. The clusters are formed on a square grid, with 3000 elementary sites on a side. Figure 2 is formed by setting the probability that a given site is occupied at $p = 0.593$, just above the critical percolation concentration p_c. The region within the small square drawn on Fig. 2 is expanded to full size in Fig. 3. Figures 2 and 3 cannot be distinguished geometrically; both obey Eq. 1 for the occupied site density on the infinite cluster (shown in white). This is an explicit example of self-similarity. Expansion or contraction of regions of Fig. 2 will look the same (that is, the site density on the infinite cluster will obey Eq. 1) as long as their size falls between the length scale limits $a \leq r \leq \xi_p$. The upper length scale ξ_p can be extracted from Fig. 2 as the size of the "largest hole" in the infinite network, or the largest cluster (shown in bright yellow). At smaller concentrations ($p < p_c$), ξ_p is the size of the largest cluster (shown in white on Fig. 1). At larger concentrations ($p > p_c$ as in Fig. 4), the infinite network encompasses more and more of the occupied sites according to Eq. 2. The correlation length can be relevant to other fractal structures. In the subsequent discussion, we retain the subscript p only when referring explicitly to the percolating network. The notation will be appropriate also for the exponents ν_p and β_p.

The fractal dimension for $d = 2$ site percolation can be obtained from Figs. 1 through 4 by drawing circles of successively increasing radii (limited, of course, always to $r < \xi_p$) and then counting the number of occupied sites of a given color lying within each circle. That number, divided by the Euclidean area, πr^2, will scale according to Eq. 1, giving D directly. The reader should be warned that there will be large fluctuations in the value found for $\rho(r)$, depending on where one chooses the origin for the circles. The magnitude of these fluctuations is a measure of the lacunarity (*1*). Averaging over a great many origins for a given circle size will result in a unique value for D. We denote this "ensemble" average by $<\ldots>$.

There are other indices besides the fractal dimension that are useful for the description of the purely static geometrical properties of fractal networks. An example is the minimum distance l traveled along the fractal network, restricted to paths between connected sites, when traversing the fractal from one point to another (for example, the minimum path length between two points on the infinite cluster for a path confined to the infinite cluster). This is termed the chemical length. It is not the same as the length L measured between the same two points with a straight edge (referred to as the Pythagorean length). The two can be related to one another by $l = L^{d_{min}}$ (*14*). In general, d_{min} is greater than or equal to 1 for fractal networks (the equality holding in the Euclidean limit, when $L \geq \xi_p$). For a percolating network in $d = 3$, d_{min} is 1.39.

Although the concept of fractal geometry is useful for the definition of static structural properties, it can also have utility when describing the dynamical properties of fractal networks. These could arise either directly, as vibrations of the atoms that make up a fractal structure, or in terms of the diffusion or flow of particles along paths constrained to fractal geometry.

The most famous example of dynamics of a fractal network originated with de Gennes (*15*), who posed the following problem. An ant is dropped onto an occupied site of the infinite cluster of a percolating network. The ant takes steps at a constant rate of elementary length a that are random in direction to adjacent (connected) sites on the infinite network. What is the (ensemble) average mean square distance that the ant travels in a time t?

This problem was solved only recently by Gefen and colleagues (*16*). The solution opened the way for a nearly complete description of the dynamics of fractal networks. They showed that the mean square distance that the ant moves in time t along the fractal network (for example, an infinite cluster) could be mapped onto the diffusion problem in Euclidean space. The price paid is the substitution of a (Euclidean) length-dependent diffusion constant [that is, $D \rightarrow D(r)$, where r is the Pythagorean length] for the conventional diffusion constant. They showed that

$$D(r) \propto r^{-\theta}, r \leq \xi_p \quad (4)$$

where the constant θ equals 0 in the Euclidean limit. Inserting Eq. 4 into the diffusion equation [$<r^2(t)> = D(r)t$] and solving for r, a scaling relation

$$<r^2(t)> \propto t^{2/(2+\theta)} \quad (5)$$

is found for a fractal network. Here, the length r is the Pythagorean distance, but diffusion has been confined to paths between connected adjacent sites on the percolating network. The value of θ has been calculated to be approximately 0.8 in $d = 2$ and 1.5 in $d = 3$ for percolating networks (*17, 18*). Diffusion is distinctly slowed down on a percolating network compared to diffusion in Euclidean space: $<r^2(t)> \propto t^{0.57}$ for $d = 3$ compared to $<r^2(t)> \propto t$, respectively.

The fracton dimension, $\bar{\bar{d}}$, first introduced by Alexander and Orbach (*19*) (they initially termed it the spectral dimension), can now be defined by rewriting Eq. 5 as

$$<r^2(t)> \propto t^{\bar{\bar{d}}/D} \quad (6)$$

where

$$\bar{\bar{d}} = 2D/(2 + \theta) \quad (7)$$

Further, diffusion on a fractal network with $\bar{\bar{d}} < 2$ is recurrent (that is, there is always a finite probability of return to the origin). This enabled Rammal and Toulouse (*20*) [and later de Gennes (*21*)] to obtain the number of distinct sites visited by the ant, $S(t)$, during its random wanderings on the fractal network in time t. De Gennes noted that the number of distinct sites contained within a sphere of radius r is proportional to r^D, so that, from Eq. 6,

$$S(t) \propto r^D \propto [(t^{\bar{\bar{d}}/D})^{1/2}]^D \equiv t^{\bar{\bar{d}}/2}, r \leq \xi_p \quad (8)$$

This result has been confirmed by many investigators by means of numerical simulations, and it is in fact a method for determining $\bar{\bar{d}}$ (*18, 22*). Mandelbrot (*1*) has observed that $\bar{\bar{d}}/2$ is the fractal codimension of the recurrences of the ant to a previously visited site. The number of distinct sites visited in time t is determined solely by $\bar{\bar{d}}$ (Eq. 8), whereas the mean square distance traveled in the same time t depends on the ratio $\bar{\bar{d}}/D$ (Eq. 6).

Earlier, Alexander and Orbach noted (*19*) that $\bar{\bar{d}}$ is approximately 4/3, to the numerical accuracy then available for percolating networks for all dimensions d greater than or equal to 2 (it is exactly 4/3 for $d \geq 6$, the dimensionality at which mean field theory is exact for percolation). They suggested that $\bar{\bar{d}} = 4/3$ may be an exact relation for all d greater than or equal to 2. This has become known as the "Alexander-Orbach conjecture." It remains rather controversial (*17*) [however, see (*18, 23*)]. It does, however, provide an excellent first approximation and may yet be found to be exact (*23*).

There is another reason for the importance of the Alexander-Orbach conjecture. Should it be exact, the exponent t, determining the dependence of the electrical conductivity σ of a random resistor network upon concentration

$$\sigma \propto (p - p_c)^t \quad (9)$$

can be related to the static geometrical exponents ν and β defined earlier in Eqs. 2 and 3. Alexander and Orbach (*19*) showed that setting $\bar{\bar{d}}$ equal to 4/3 for a percolating network results in

$$t_p = ½[\nu_p(3d - 4) - \beta_p] \quad (10)$$

As noted before, ν_p is 4/3 and β_p is 5/36 in $d = 2$ for a percolating network, leading to $t_p = 91/72(\simeq 1.264)$ in $d = 2$. Recent numerical simulations give values for t_p that are closer to 1.3, or a value for $\bar{\bar{d}}$ of about 1.3 (*17*). However, Rammal and colleagues (*18*) find $\bar{\bar{d}} = 4/3$ to within numerical accuracy. A recent series expansion calculation (*23*) finds $\bar{\bar{d}} = 1.334 \pm 0.007$.

The fracton dimensionality may be most useful for the description of the vibrational dynamics of fractal networks. An example is the percolating network with each occupied site replaced by an identical mass and each bond replaced by an identical spring. Montroll (*24*) showed three decades ago that, if one could solve the problem of diffusion on such a network [here, the ant problem posed by de Gennes (*15*)], then the vibration problem of the geometrically equivalent network was automatically solved [for a force between atoms of the "scalar" type (*25*)].

It has already been noted that the geometry changes from fractal at short length scales ($a \leq r \leq \xi$) to Euclidean at large length scales ($r \geq \xi$). Short length scales mean high-frequency vibrations. These were termed fractons by Alexander and Orbach (*19*). Long length scales mean low-frequency vibrations. These are conventionally termed phonons (*26*). Thus, fractons are distinguished from phonons according to whether their wavelengths are short or long compared to ξ, respectively. We can associate the length ξ with a characteristic (or "crossover") frequency ω_c according to the relationship

$$\omega_c \propto \xi^{-D/\bar{\bar{d}}} \quad (11)$$

Thus, fracton frequencies are greater than ω_c, and phonon frequencies are less than ω_c.

The vibrational energy density of states $N(\omega)$ associated with a vibrating fractal network can be obtained easily. This quantity is important, for example, for the specific heat and the thermal conductivity of a fractal network. It is known (*26*) that $N(\omega)$ is proportional to $\omega^{\bar{d}-1}$ for Euclidean structures in the Debye ($\omega < \omega_c$) regime. Alexander and Orbach (*19*) showed that $N(\omega)$ is proportional to $\omega^{\bar{\bar{d}}-1}$ in the fractal ($\omega > \omega_c$) regime. The form of $N(\omega)$ as a function of ω therefore depends on the excitation frequency of the fractal network. Thus, $N(\omega)$ is proportional to ω^2 for a $d = 3$ Euclidean structure in the Debye limit, whereas $N(\omega)$ is proportional to $\omega^{1/3}$ (from the Alexander-Orbach conjecture) for any $d \geq 2$ percolating network. This difference of behavior has been verified numerically (*27*) for percolating networks in $d = 2$ and $d = 3$. The precise shape of $N(\omega)$ in the vicinity of ω_c remains to be determined. A recent scaling analysis (*28*) suggests a bumplike structure in the vicinity of ω_c (Fig. 5).

Amorphous solids may exhibit fractal geometry at short length scales (*29*). Their vibrational density of states should follow the general outlines of Fig. 5. Recent inelastic neutron diffraction experiments (*30*) on vitreous silica (Fig. 6) exhibit $N(\omega)$ down to remarkably low vibrational energies: $\hbar\omega/k_B = 1.6$ K(k_B is Boltzmann's constant). The comparison with Fig. 5 is striking, but it should be stated that the shape of $N(\omega)$ has been attributed to a specific collective rotation of SiO_2 tetrahedra (perhaps the physical description of a vibrational excitation on a fractal network) (*30*).

In addition to describing the energy of vibrational excitations,

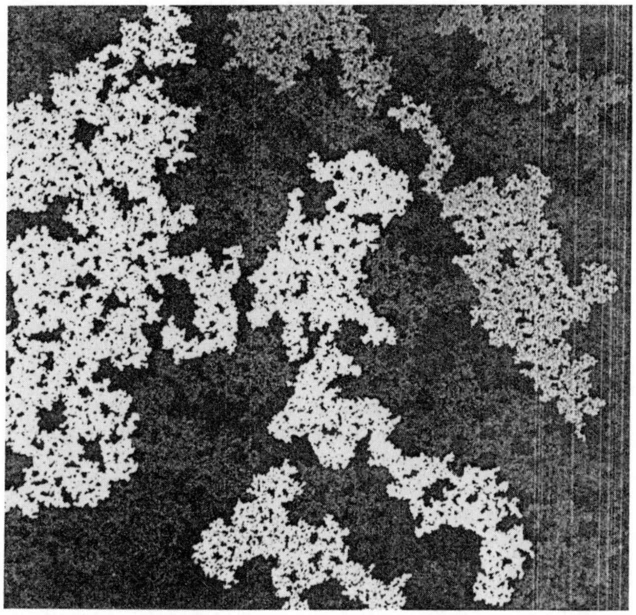

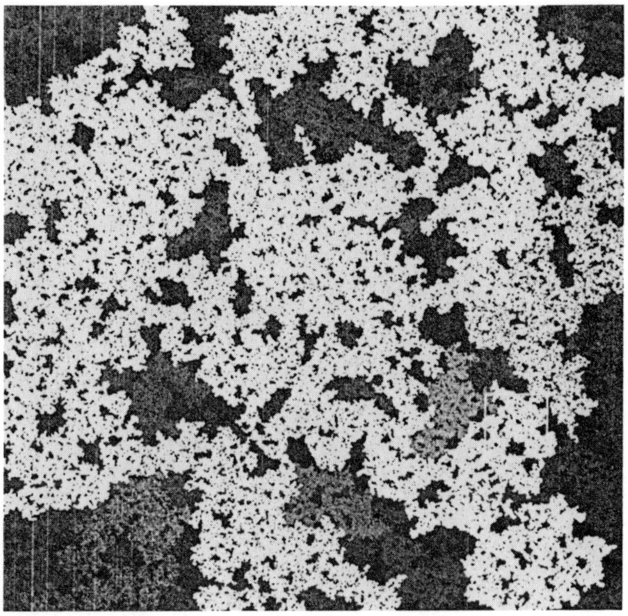

Fig. 1 (left). A percolation network, for sites at concentration $p = 59.00$ percent, occupied at random on a square grid of 3000 sites on a side, derived from periodic boundary conditions. Sites are considered connected if they are adjacent along a vertical or horizontal direction, not along the diagonals. The arrays of connected sites are clusters. The critical concentration for the largest cluster to span the grid from one side to another is 59.28 percent, in $d = 2$, so that this network is below the critical concentration. The size of the largest cluster (shown in white) is a measure of the percolation correlation length ξ_p at $p = 59.00$ percent. The structure is fractal for length scales $a \leq r \leq \xi_p$, where a is the fundamental grid size. [Reprinted with permission from R. F. Voss, IBM Thomas J. Watson Research Laboratories]

Fig. 2 (right). The same construction as in Fig. 1, but at $p = 59.30$ percent, just above the critical concentration for site percolation, $p_c = 59.28$ percent, in $d = 2$. The white cluster is now an infinite cluster (spanning the grid from one side to the other), and the percolation correlation length ξ_p is of the order of the next largest cluster (colored bright yellow). The structure is fractal for length scales $a \leq r \leq \xi_p$. [Reprinted with permission from R. F. Voss, IBM Thomas J. Watson Research Laboratories]

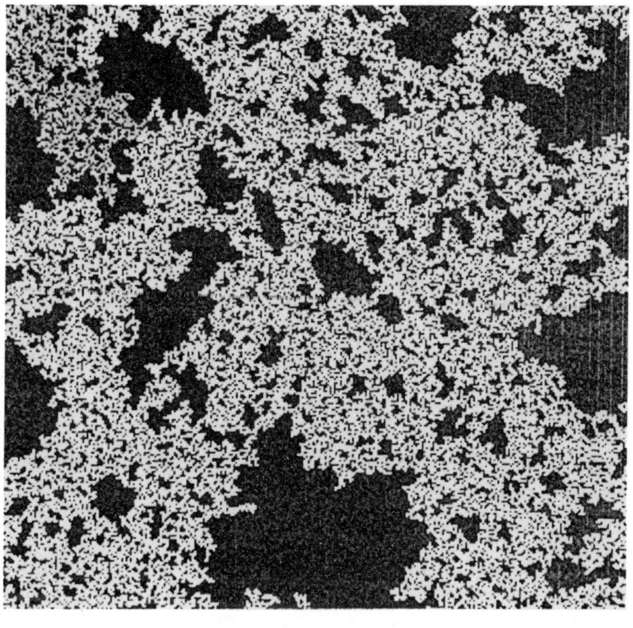

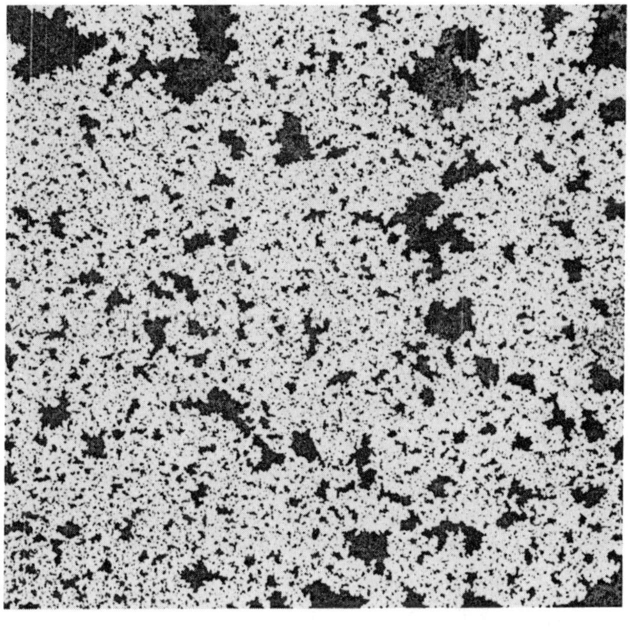

Fig. 3 (left). The expansion to full size of the small square in Fig. 2 (site percolation on a square grid at $p = 59.3$ percent). The grain is sufficiently large to observe the microscopic grid size. This is the lower limit a to fractal geometry. The average geometrical structure of the infinite cluster (shown in white) is the same as in Fig. 2. This is an illustration of self-similar geometry. [Reprinted with permission from R. F. Voss, IBM Thomas J. Watson Research Laboratories]

Fig. 4 (right). The same construction as in Fig. 1, but at $p = 59.60$ percent, above the critical concentration for site percolation, $p_c = 59.28$ percent, in $d = 2$. Again, ξ_p is the size of the largest hole (or the bright yellow cluster), and the structure is fractal for $r \leq \xi_p$. [Reprinted with permission from R. F. Voss, IBM Thomas J. Watson Research Laboratories]

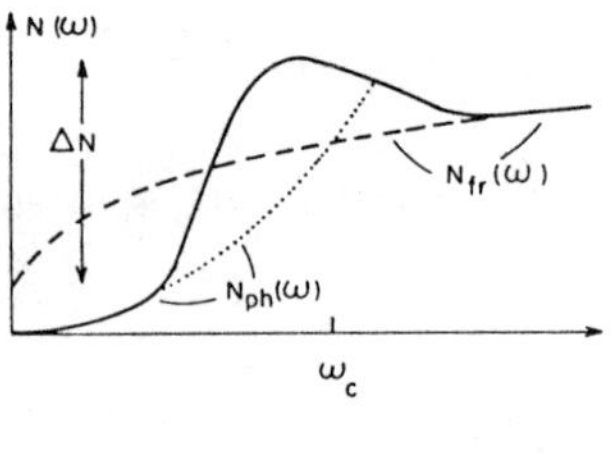

Fig. 5. A sketch of the density of vibrational states for a fractal network [the solid line, denoted $N_{fr}(\omega)$] plotted against frequency. The extrapolation of the low-frequency (long length scale) phonon density of states, $N_{ph}(\omega)$, is shown by a dotted line. The extrapolation of the high-frequency (short length scale) fracton density of states, $N_{fr}(\omega)$, is shown by the dashed line. [From (*28*)]

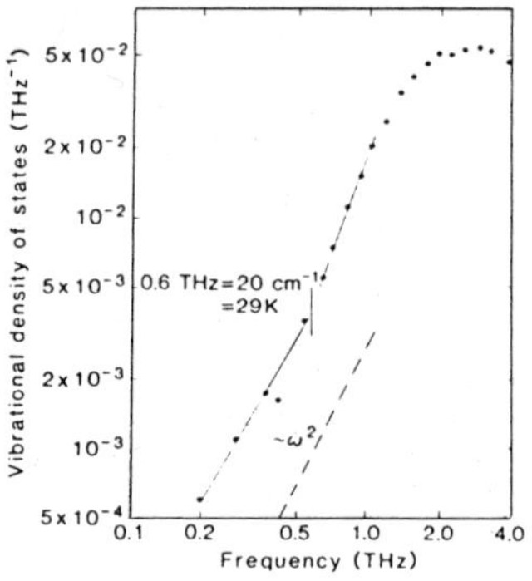

Fig. 6. A plot of the vibrational density of states for vitrous silica at 290 K. A line has been drawn at an energy that could reasonably be identified with the crossover from phonon to fracton excitations. Note the similarity to the predictions of Fig. 5. [Data replotted by U. Buchenau *et al.* (*30*)]

fractal geometry can also be useful for determination of the spatial extent of the vibrational wave function. This is particularly useful for random systems where explicit calculation of the wave function would be very difficult. There are two extremes: extended and localized. The former is the condition for homogeneous structures (weak disorder), where a plane wave is a useful first approximation. The latter is relevant to sufficiently strong disorder, where the wave function falls off exponentially (or more rapidly) with distance. Rammal and Toulouse (*20*) applied the scaling theory of localization (*31*) to excitations on a fractal network. The key parameter in that theory is the exponent β which governs the proportionality of the electrical conductance Σ for a random resistor network to the distance over which Σ is measured:

$$\Sigma(r) \propto r^{\beta} \tag{12}$$

For zero or negative β, the wave functions are localized. For positive β, they are extended. When applied to fractal networks, Rammal and Toulouse showed (*20*)

$$\beta = (D/\bar{\bar{d}})(\bar{\bar{d}} - 2) \tag{13}$$

For percolating networks, the Alexander-Orbach conjectures gives $\bar{\bar{d}} = 4/3$, leading to a strongly negative value for β. This results in localized fracton states.

The spatial extent of the localized fracton states, λ, depends on their excitation energy ω (*19*):

$$\lambda(\omega) \propto \omega^{-\bar{\bar{d}}/D} \tag{14}$$

This "dispersion law" depends only on the fracton and fractal dimensions. It is an example of how fractal geometry can simplify complex relations.

The form of the localized fracton wave function has been hypothesized (*32*) to be of the form

$$\phi(\omega) \propto (1/r^{d-D})\,[\lambda(\omega)]^{-D/2} \exp\{-\tfrac{1}{2}[r/\lambda(\omega)]^{d_{\phi}}\} \tag{15}$$

This is known (*33*) as "super-localization" because the exponent d_{ϕ} is bounded by unity and d_{min} and so in general is greater than unity. However, little is known about the precise shape of the fracton wave function in random structures (even for the well-studied percolating network), so that Eq. 15 should be regarded as a first approximation.

The localized character of fracton excitations can have profound implications. Consider, for example, the thermal conductivity of a fractal network. The long length scale (low-energy) vibrational excitations are phonons that, in $d = 3$, can be extended (because in Euclidean space $\bar{\bar{d}} = D = d$ and $\beta > 0$ in Eq. 13). The phonons are therefore able to carry heat from one spatial region to another. However, as the temperature increases, higher energy vibrations will be excited. Their energy is denoted by $\hbar\omega$, where thermal excitation means $\hbar\omega \sim k_{B}T$. When ω is approximately equal to ω_c, higher temperatures will excite fractons and not phonons (Fig. 5). However, fractons cannot carry heat from one spatial region to another because they are localized for $\bar{\bar{d}} < 2$. This means that $k_{B}T$ will be much greater than the energy of those vibrational states that can carry heat (the phonons). Assuming a temperature-independent mean free path for the phonons then implies a thermal conductivity independent of temperature. Such a "plateau" in the thermal conductivity is common to nearly all amorphous solids (*34*). However, as the temperature is further increased, eventually the thermal conductivity is seen to increase. The origin of this increase has been attributed to thermal conductance arising from phonon-assisted fracton "hopping" (*35*) as a consequence of vibrational anharmonicity. The thermal conductivity is predicted (*35*) to increase linearly with increasing temperature, which is consistent with observation (*34*). Other interpretations (*36*) have also been offered for this behavior. They involve scattering processes off of "two-level systems" common to amorphous materials. The fracton-hopping picture is a "cleaner" explanation, but the matter is far from settled.

There are further implications of fracton localization. Excited states of localized electronic centers can relax nonradiatively by emitting one or more lattice vibrations (such processes are central to laser operation). Localized paramagnetic centers or nuclei relax toward thermal equilibrium (the so-called T_1 spin-lattice relaxation process) by inelastic scattering of lattice vibrations. These processes involve the interaction of electrons and lattice vibrations. Although the theory is well developed for phonons, an equivalent development has only recently been accomplished for fractons (*37*).

The simplest example is the one-vibrational quanta direct relaxation process (*38*). A localized electronic center relaxes to a lower energy state by emitting a lattice vibration, the energy of which is equal to the change in electronic energy. However, if the vibrational state is also localized, the two may lie farther away from one another than the localization length of the vibrational state (given for fractons by Eq. 14). The strength of the electron-fracton coupling for this particular pair of electron and fracton states would then be very small, inhibiting the electronic relaxation process. In general, there will be a distribution of distances between the electronic sites and the fractons that conserves energy in the relaxation process. This will result in an electronic relaxation rate that will vary from site to site. The probability density for the relaxation rate can be evaluated explicitly (*37*). The probability $P(t)$ to find the electron in its initial state after time t will vary at long times as

$$P_{direct}(t) \propto t^{-(ln\ t)^{x}} \tag{16}$$

where $x = (D/d_{\phi}) - 1$. For localized phonons, $D = d$ and $d_{\phi} = 1$, so that $x = 2$ in $d = 3$.

The direct-process electronic time relaxation profile (Eq. 16) is slower than exponential but faster than any power law. It is a direct consequence of localized vibrational states interacting with localized electronic sites. Extended vibrational states (phonons) will give rise to a single relaxation rate and an exponential time dependence for $P_{direct}(t)$ in place of Eq. 16 (*38*).

Electronic (or nuclear) spin-lattice relaxation transitions involve relatively small changes in energy. The density of vibrational states at these small energies is too small to make the direct relaxation process effective. Rather, two vibrational states are involved, with the change in electronic (or nuclear) energy equaling the difference in the vibrational state energies. This is known as the Raman relaxation process (*38*). Extended vibrational states would again generate an exponential time dependence for $P(t)$. The temperature dependence varies as T^9 or T^7, depending on whether the electronic transition takes place between time-reversed states of half-integral spin (termed Kramers transitions) or not (termed non-Kramers transitions). Localized vibrational states again profoundly affect the time dependence of the Raman relaxation process. In the "near" long-time regime, the probability $P(t)$ to find the electron in its initial state after time t will vary as

$$P_{\text{Raman}}(t) \propto \exp[-t^{1/(a-2)}] \tag{17}$$

where $a = 4\bar{\bar{d}}(d_\phi/D) + 2\bar{\bar{d}}$ or $4\bar{\bar{d}}(d_\phi/D) + 2\bar{\bar{d}} - 2$ for Kramers or non-Kramers transitions, respectively. In the "far" long-time regime,

$$P_{\text{Raman}}(t) \propto t^{-(\ln t)^{2D/d_\phi}} \tag{18}$$

which is similar to the time dependence of the direct process.

The calculation of the temperature dependence of the Raman relaxation process is greatly complicated by the nonexponential time dependence of both Eqs. 17 and 18. However, a simplification results when the (dipole-dipole or exchange) coupling between spins is sufficiently strong that the spins relax as a unit (that is, strong cross-relaxation). The time profile $P(t)$ is then exponential and is given by an average relaxation rate (written as an inverse relaxation time $1/T_1$), proportional to

$$(1/T_1)_{\text{Raman}} \propto T^{2\bar{\bar{d}}[1 + 2(d_\phi/D)] - 1} \qquad \text{(Kramers)} \tag{19a}$$

and

$$(1/T_1)_{\text{Raman}} \propto T^{2\bar{\bar{d}}[1 + 2(d_\phi/D)] - 3} \qquad \text{(non-Kramers)} \tag{19b}$$

These temperature dependences become $T^{2.63}$ and $T^{4.63}$, respectively, for a percolating network upon setting d_ϕ equal to d_{min}. Extended vibrational states would generate temperature dependences of T^9 and T^7, respectively. Experimental evidence for an anomalous temperature dependence for the Raman relaxation process has been obtained (*39*) for low-spin iron in large biological molecules.

These calculations demonstrate that local probes embedded in fractal networks should exhibit substantially different relaxation time profiles and temperature dependences as compared to Euclidean embeddings. Use of localized electronic probes in amorphous materials may be a direct test of the hypothesis that fractal geometry is a useful geometrical description at short length scales. The consequences of a fractal geometry are manifest through Eqs. 16 through 19.

REFERENCES AND NOTES

1. B. B. Mandelbrot, *The Fractal Geometry of Nature* (Freeman, New York, 1983); *Ann. Israel Phys. Soc.* **5**, 59 (1983); *J. Stat. Phys.* **34**, 895 (1984).
2. *J. Stat. Phys.* **36**, nos. 5 and 6 (1984); F. Family and D. P. Landau, Eds., *Kinetics of Aggregation and Gelation* (Elsevier/North-Holland, Amsterdam, 1984); *Scaling Phenomena in Disordered Systems* (NATO Advanced Study Institute, Geilo, Norway, April 1985) (Plenum, New York, 1985); H. E. Stanley, Ed., *Growth and Form* (Nijhoff, Amsterdam, 1985); L. Pietronero, Ed., *Fractals in Physics* (Elsevier/North-Holland, Amsterdam, 1986).
3. Cab-O-Sil is a trademark of the Cabot Corporation; Alfasil can be found as Stock Number D89376 (Alfa Catalog, 1981).
4. S. K. Sinha, T. Freltoft, J. Kjems, in *Kinetics of Aggregation and Gelation*, F. Family and D. P. Landau, Eds. (Elsevier/North-Holland, Amsterdam, 1984), p. 87.
5. A. Aharony, *J. Stat. Phys.* **34**, 931 (1984); D. Stauffer, *Introduction to Percolation Theory* (Taylor and Francis, London, 1985).
6. D. Stauffer, A. Coniglio, M. Adam, *Adv. Poly. Sci.* **44**, 103 (1982).
7. P. G. de Gennes, *Scaling Concepts in Polymer Physics* (Cornell Univ. Press, Ithaca, NY, 1979).
8. M. F. Thorpe, *J. Non-Cryst. Solids* **57**, 3155 (1983); H. He and M. F. Thorpe, *Phys. Rev. Lett.* **54**, 2107 (1985).
9. P.-G. de Gennes, *J. Phys. (Paris) Colloq.* **41**, C3/17 (1980); B. I. Halperin, S. Feng, P. N. Sen, *Phys. Rev. Lett.* **54**, 2391 (1985).
10. M. P. M. den Nijs, *J. Phys. A* **12**, 1857 (1979).
11. B. Nienhuis, E. K. Riedel, M. Schick, *ibid.* **13**, L189 (1980).
12. R. F. Voss, R. B. Laibowitz, E. I. Allessandrini, *Phys. Rev. Lett.* **49**, 1441 (1982); A. Kapitulnik and G. Deutscher, *ibid.*, p. 144.
13. Courtesy of R. F. Voss.
14. K. M. Middlemiss, S. G. Whittington, D. S. Gaunt, *J. Phys. A* **13**, 1835 (1980); R. Pike and H. E. Stanley, *ibid.* **14**, L169 (1981); D. C. Hong and H. E. Stanley, *ibid.* **16**, L475 (1983); *ibid.*, p. L525; H. J. Herman, D. C. Hong, H. E. Stanley, *ibid.* **17**, L261 (1984); J. Vannimenus, J. P. Nodal, C. Martin, *ibid.*, p. L351; S. Havlin and R. Nossal, *ibid.*, p. L425; S. Havlin, Z. V. Djordjevic, I. Majid, H. E. Stanley, G. H. Weiss, *Phys. Rev. Lett.* **53**, 178 (1984).
15. P. G. de Gennes, *Recherche* **7**, 919 (1976).
16. Y. Gefen, A. Aharony, S. Alexander, *Phys. Rev. Lett.* **50**, 77 (1973).
17. D. C. Hong, S. Havlin, H. J. Hermann, H. E. Stanley, *Phys. Rev. B* **30**, 4083 (1984); J. G. Zabolitzky, *ibid.*, p. 4077; H. J. Herrmann, B. Derrida, J. Vannimenus, *ibid.*, p. 4080; C. J. Lobb and D. J. Frank, *ibid.*, p. 4090.
18. R. Rammal, J. C. Angles d'Auriac, A. Benoit, *ibid.*, p. 4087.
19. S. Alexander and R. Orbach, *J. Phys. (Paris) Lett.* **43**, L-625 (1982).
20. R. Rammal and G. Toulouse, *ibid.* **44**, L-13 (1983).
21. P.-G. de Gennes, *C. R. Acad. Sci. Ser. B* **296**, 881 (1983).
22. P. Argyrakis and R. Kopelman, *Phys. Rev. B* **29**, 511 (1984).
23. J. W. Essam and F. M. Bhatti, in preparation.
24. E. W. Montroll, in *Proceedings of the Third Berkeley Symposium on Mathematical Statistics and Probability* (Univ. of California Press, Berkeley, 1955), p. 209.
25. S. Alexander, *J. Phys. (Paris)* **45**, 1939 (1984).
26. C. Kittel, *Introduction to Solid State Physics* (Wiley, New York, ed. 5, 1976), p. 107.
27. I. Webman and G. Grest, *Phys. Rev. B* **31**, 1690 (1985); *J. Phys. (Paris) Lett.* **45**, L-1155 (1984).
28. A. Aharony, S. Alexander, O. Entin-Wohlman, R. Orbach, *Phys. Rev. B* **31**, 2565 (1985); see also the earlier work of P. F. Tua, S. J. Putterman, and R. Orbach [*Phys. Lett.* **98A**, 357 (1983)].
29. S. Alexander, C. Laermans, R. Orbach, H. M. Rosenberg, *Phys. Rev. B* **28**, 4615 (1983).
30. U. Buchenau, N. Nucker, A. J. Dianoux, *Phys. Rev. Lett.* **53**, 2316 (1984).
31. E. Abrahams, P. W. Anderson, D. C. Licciardello, T. V. Ramakrishnan, *ibid.* **42**, 673 (1979).
32. S. Alexander, O. Entin-Wohlman, R. Orbach, *Phys. Rev. B* **32**, 6447 (1985).
33. Y.-E. Levy and B. Souillard, unpublished results.
34. R. C. Zeller and R. O. Pohl, *Phys. Rev. B* **4**, 2029 (1971); M. P. Zeitlin and A. C. Anderson, *ibid.* **12**, 4475 (1975).
35. S. Alexander, O. Entin-Wohlman, R. Orbach, unpublished results.
36. V. G. Karpov and D. A. Parshin, *J. Exp. Theor. Phys. Lett.* **38**, 648 (1983).
37. S. Alexander, O. Entin-Wohlman, R. Orbach, *J. Phys. (Paris) Lett.* **46**, L-549 (1985); *ibid.*, p. L-555.
38. R. Orbach and H. Stapleton, in *Electronic Paramagnetic Resonance*, S. Geschwind, Ed. (Plenum, New York, 1972), p. 121.
39. H. J. Stapleton, J. P. Allen, C. P. Flynn, D. G. Stinson, S. Kurtz, *Phys. Rev. Lett.* **45**, 1456 (1980); J. P. Allen *et al.*, *Biophys. J.* **38**, 299 (1982).

40. I thank D. O. Raphael for critical reading of the manuscript. Supported by NSF grant DMR 84-12898.

VOLUME 50, NUMBER 1 PHYSICAL REVIEW LETTERS 3 JANUARY 1983

lished.

[5]R. Eichele, H. Seifert, and R. P. Huebner, Appl. Phys. Lett. 38, 383 (1981).

[6]P. L. Stöhr, Appl. Phys. Lett. 39, 828 (1981).

[7]C. C. Chi, M. M. T. Loy, and D. C. Cronemeyer, Appl. Phys. Lett. 40, 437 (1982).

Anomalous Diffusion on Percolating Clusters

Yuval Gefen and Amnon Aharony
Department of Physics and Astronomy, Tel Aviv University, Ramat Aviv, Israel

and

Shlomo Alexander
The Racah Institute of Physics, The Hebrew University, Jerusalem, Israel

(Received 7 September 1982)

The mean square distance reached after t random-walk steps on a percolating cluster is shown to behave as $t^{2/(2+\theta)}$ (instead of t) for short times (for which the cluster is seen to be self-similar). The exponent θ is related to that describing the dc conductivity near percolation. Averaging over all clusters yields the ac conductivity and dielectric constant near percolation in the high-frequency limit, when the polarization of the medium is unimportant.

PACS numbers: 71.30.+h, 05.50.+q

Much of the current interest in the properties of dilute systems concentrates on the vicinity of the percolation threshold, p_c.[1,2] As the concentration p of, e.g., conducting links, in a metal-insulator alloy, approaches p_c, the percolation correlation (i.e., the pair connectedness) length diverges as $\xi \sim |p-p_c|^{-\nu}$. For $p>p_c$, the probability to belong to the infinite cluster is $P_\infty(p) \propto (p-p_c)^\beta \propto \xi^{-\beta/\nu}$ and the dc conductivity approaches zero as $\sigma_{dc} \propto (p-p_c)^\mu \propto \xi^{-\mu/\nu}$. Both P_∞ and σ_{dc} vanish for $p<p_c$. In a superconductor–normal-metal alloy, one has $\sigma_{dc} \propto (p_c-p)^{-s}$ for $p<p_c$. The exponent s has recently also been predicted to describe the divergence of the dielectric constant of a metal-insulator alloy as p_c is approached from below.[3,4]

All these results represent *macroscopic* averages, taken over samples which are *large compared to* ξ. On such large length scales, the samples look *homogeneous*. However, both Monte Carlo simulations[2] and experiments[5] show that the geometry of the percolating cluster is not homogeneous for length scales L which are in the range $a \ll L \ll \xi$ (a is the microscopic lattice distance). In this range the clusters are *self-similar*, and the number of links which belong to a given cluster (whose size is larger than L) within a volume L^d (in d dimensions) varies as L^{d_f}, with the fractal dimensionality $d_f = d - \beta/\nu$.[2,6,7] Note that this property is expected of both the finite and the infinite clusters. On "anomalous" length scales L, with $a \ll L \ll \xi$, one expects all physical measurements to reflect this self-similarity and to yield results which behave as powers of L, *independent* of ξ.[7] Bulk properties are then found by a juxtaposition of pieces of size ξ. The aim of the present Letter is to use scaling considerations for deducing these "anomalous" properties and the crossover at $L \approx \xi$.[8]

Our main results concern the problem of diffusion on random networks ("the unbiased ant in the labyrinth").[9-12] Consider a single percolation cluster, containing s sites within a linear range of order $r_s \propto s^{1/d_f}$. If the random walker (the "ant") is put on this cluster, and if the distance it travels after t time steps, $r(t)$, satisfies $a \ll r(t) \ll (r_s, \xi)$, then self-similarity implies that

$$\langle r^2(t)\rangle_s = A^2 t^{2/(2+\theta)}, \tag{1}$$

for *both* $p>p_c$ and $p<p_c$. One may define an *effective dimensionality of the random walk*,[13] d_{rw}, via $t \propto r^{d_{rw}}$. Our result implies that $d_{rw} = 2+\theta$, instead of the usual Fickian value $d_{rw}=2$ (i.e., $\langle r^2\rangle \propto t$). (In the notation of Ref. 6, Chaps. 27–30, $2+\theta = 1/H$.) The latter is recovered only for $p>p_c$ and $r(t) \gg \xi$. The exponent θ is given by $\theta = (\mu-\beta)/\nu$. Using known estimates for μ, β,

VOLUME 50, NUMBER 1 PHYSICAL REVIEW LETTERS 3 JANUARY 1983

and ν,[1] one has $\theta \simeq 0$, 0.8, 1.5, and 2.7 for $d=1$, 2, 3, and 4. The mean-field result, holding for $d \geqslant 6$, is $\theta = 4$.

The derivation of Eq. (1) is based on a scaling description of the diffusion coefficient, $D_s = \frac{1}{2} \times d\langle r^2(t)\rangle_s/dt$. For $p > p_c$, when $\langle r^2(t)\rangle_s \gg \xi^2$, the term linear in t in $\langle r^2(t)\rangle_s$ results from motion on the infinite cluster. The diffusion constant on this cluster, D_∞, is related to the dc conductivity σ_{dc} via the Einstein relation, $\sigma_{dc} = e^2 n D_\infty / kT$. On large length scales, only the electrons on the infinite cluster contribute to the conductivity. Their density n is therefore proportional to P_∞, and we find that $D_\infty \propto (p-p_c)^{\mu-\beta} \propto \xi^{-\theta}$. Thus, $\langle r^2(t)\rangle_\infty \propto t\xi^{-\theta}$. In general, D_∞ should depend on the length scale L via the scaling form $D_\infty(L) = \xi^{-\theta} f(L/\xi)$. For "anomalous" scales $a \ll L \ll \xi$, we expect D_∞ to become independent of ξ and hence $f(x) \propto x^{-\theta}$ for $x \ll 1$, i.e., $D_\infty(L) \propto L^{-\theta}$.[14] On these scales, the same result should apply to diffusion of *any* single cluster with $r_s \gg L$. Integrating the equation $dr^2/dt \propto r^{-\theta}$ we recover Eq. (1).

For $r(t) < \xi$, the result (1) applies only to $r(t) \simeq At^{1/(2+\theta)} \ll r_s$. For times $t \gg (r_s/A)^{2+\theta}$ we expect that $\langle r^2(t)\rangle_s \to r_s^2$. All of these results may be derived as special limits of the general scaling form

$$\langle r^2(t)\rangle_s = t\xi^{-\theta} f_\pm(t/\xi^{2+\theta}, r_s/\xi), \tag{2}$$

where $+(-)$ denotes $p > p_c$ $(p < p_c)$. For $r_s \gg \xi$, this form is similar to that of Straley[12] (his y is our $2+\theta$), and one can follow his derivation of our result for θ by setting $f_+(\infty, \infty)$ equal to a constant. Our new "anomalous" result (1) follows if $f_\pm(x, \infty) \propto x^{-\theta(2+\theta)}$ for $x \ll 1$.

We now turn to an average over all the clusters. The distribution of the finite ones may be approximated by the form[1] $n_s \propto s^{-\tau}$ for $r_s < \xi$, and $n_s \simeq 0$ for $r_s > \xi$.[15] The average diffusion over all the finite clusters is thus given by $\langle r^2(t)\rangle \propto \int_1^{\xi^{d_f}} ds\, s n_s \times \langle r^2(t)\rangle_s$. (We now assume that all lattice sites have the same probability to be selected as the starting point.) Substituting $\langle r^2(t)\rangle_s = r_s^2$ for $s < [At^{1/(2+\theta)}]^{d_f}$, and Eq. (1) for the remaining range (as well as for the infinite cluster), we find that the *average* diffusion coefficient at time t, with $L = At^{1/(2+\theta)} < \xi$, has the form

$$\bar{D}(L) = \tfrac{1}{2} d\langle r^2\rangle/dt$$
$$= C_1 L^{-\mu/\nu}[1-(L/\xi)^{\beta/\nu}] + C_2 P_\infty(p) L^{-\theta}. \tag{3}$$

For $p < p_c$ one has $P_\infty = 0$, and $\bar{D}(L)$ is proportional to the (size-dependent) dc conductivity $\sigma_{dc}(L) \propto L^{-\mu/\nu}$. This is expected from the Einstein relation, since the *total* electron density is noncritical. Substituting $L = At^{1/(2+\theta)}$, and integrating the diffusion equation, we now find that

$$\langle r^2(t)\rangle = t\xi^{-\mu/\nu} \bar{f}_\pm(t/\xi^{2+\theta}). \tag{4}$$

Averaging over all clusters thus replaces θ in the prefactor of Eq. (2) by μ/ν. Clearly $\bar{f}_+(\infty)$ is a constant, and $\bar{f}_\pm(x) \propto x^{-\mu/\nu(2+\theta)}$ for $x \ll 1$. The single-cluster rule (1) is thus replaced by $\langle r^2(t)\rangle \propto t^{(2-\beta/\nu)/(2+\theta)}$ for $t \ll (\xi/A)^{2+\theta}$. Note that we also require that $\tau_0 \ll t$, where τ_0 is some microscopic time (related, e.g., to the length a).

For $p < p_c$ we find that $\bar{f}_-(x) \propto 1/x$ for $x \gg 1$, i.e., $\langle r^2(t)\rangle \propto \xi^{2-\beta/\nu}$ for $t \gg (\xi/A)^{2+\theta}$. This is indeed the expected mean square radius of the clusters, $\langle r_s^2\rangle$.[1] Crossovers to $t\xi^{-\mu/\nu}$ $(p > p_c)$ and $\xi^{2-\beta/\nu}$ $(p < p_c)$, occurring at time $t \propto \xi^{2+\theta}$, were indeed observed in the numerical studies of Refs. 10 and 11. However, the results for intermediate length scales were not analyzed with the form (1). We hope that this Letter will stimulate such detailed analyses.[16] This may turn out to be a very accurate way to determine the exponent μ, for which earlier estimates scatter widely.

The anomalous diffusion which we find at short times must have implications for the ac conductivity at high frequencies, for which the charge carriers move only on "anomalous" length scales. Since our diffusion model allows the random walker to move only within the single cluster on which it was initially put, an equivalent electric circuit should contain finite resistors between occupied sites, which are connected to the ground via unit capacitors.[17] The average wave-vector- and frequency-dependent ac conductivity of such a circuit will be proportional to the Fourier transform of our $\bar{D}$, and will have the scaling form

$$\sigma_{ac}(q, \omega) = \xi^{-\mu/\nu} S_\pm(q\xi, \omega\xi^{2+\theta}). \tag{5}$$

One thus predicts that the anomalous diffusion will yield $\sigma_{ac} \propto q^{t/\nu}$ if $q\xi \gg 1$, $\omega/(Aq)^{2+\theta} \ll 1$ (region A), while $\sigma_{ac} \propto \omega^{\mu/\nu(2+\theta)}$ if $\omega\xi^{2+\theta} \gg 1$, $\omega/(Aq)^{2+\theta} \gg 1$ (region B), for *both* $p > p_c$ and $p < p_c$. The borderline between the two regions, $\omega \simeq (Aq)^{2+\theta}$, represents the line on which a distance $1/q$ is traveled during the time $1/\omega$. The bulk behavior of σ_{ac} in region A (B) is found by a juxtaposition of boxes of size $1/q$ $(A/\omega^{1/(2+\theta)})$. Regions A and B have the trivial upper microscopic limits $\omega < 1/\tau_0$ and $q < 1/a$.

With use of the Kramers-Kronig relation $\epsilon - 1 \propto P\int d\omega' \sigma_{ac}(q, \omega')/\omega'(\omega' - \omega)$, Eq. (5) implies that the average dielectric constant scales as

$\epsilon(q,\omega)-1\propto\xi^{2-\beta/\nu}E_\pm(q\xi,\omega\xi^{2+\theta})$, yielding $\epsilon-1\propto q^{(\beta/\nu)-2}$ in region A and $\epsilon-1\propto\omega^{(\beta-2\nu)/\nu(2+\theta)}$ in region B, both above and below p_c.

We now turn to the static limit, $q\xi\ll1$, $\omega\xi^{2+\theta}\ll1$. For $p>p_c$, this limit yields the macroscopic metallic behavior, $\sigma\propto\bar{D}\omega^2/[\omega^2+(\bar{D}q^2)^2]$ and $(\epsilon-1)\propto\bar{D}^2q^2/[\omega^2+(\bar{D}q^2)^2]$. For $p<p_c$, our scaling form predicts that ϵ diverges as $\epsilon(0,0)\propto\xi^{2-\beta/\nu}$.

Previous calculations of σ_{ac} used a completely different approach. The average conductivity of a mixture with two conductances σ_1 and σ_2 was written in the scaling form[3,18] $\sigma=\sigma_1|p-p_c|^\mu\tilde{S}_\pm(\sigma_2/\sigma_1|p-p_c|^{s+\mu})$. It was then assumed that this could be extended analytically by setting[3,4] $\sigma_2=i\omega\epsilon_2/4\pi$. This yields the scaled variable $\omega/|p-p_c|^{s+\mu}$ and the static dielectric constant $\epsilon\propto(p-p_c)^{-s}$. The two approaches give the same results only if we identify $s=2\nu-\beta$. Indeed, this value of s was found by Stephen[17] for the circuit which is equivalent to our diffusion problem. The result generally holds for mean-field calculations, which yield[19] $s=2\nu-\beta=0$, and for the effective medium theory,[20] in which $s=2\nu-\beta=1$. However, the equality $s=2\nu-\beta$ definitely does not hold at $d=3$ [$s\simeq0.6$,[19] $2\nu-\beta\simeq1.3$ (Ref. 1)] or at $d=2$ [$s=\mu\simeq1.2$,[19] $2\nu-\beta=\frac{91}{36}$ (Ref. 1)].

The discrepancy between the two models is not really surprising. Since we start from the diffusion equation, our circuit *does not contain capacitances on bonds connecting the different clusters.* The polarization of the medium within which these clusters are embedded (represented by ϵ_2 in the alternative scaling approach) is taken to be zero. Instead, we take account of the *anomalous diffusion within the clusters,* which was ignored in all the earlier studies. In general, the polarization of the medium will reduce the effective dielectric constant, and turn it from our $\xi^{2-\beta/\nu}$ into $\xi^{s/\nu}$. The equality $s=2\nu-\beta$, found within mean-field theory, which is valid at dimensionalities $d>6$, implies that the polarization of the medium is not singular for $d>6$. This is to be expected, since at high d the distances among clusters are large (p_c is low) and the electrostatic interaction decays with distance as $1/r^{d-2}$.

Note that our average static dielectric constant at $p<p_c$ behaves as $\epsilon-1\propto\xi^{2-\beta/\nu}\propto\langle r_s^2\rangle$. Each site on a cluster of size s thus has a polarization of order $P\propto r_s^2E$ in an external field E. Such behavior can be directly obtained from linear response theory, in which the contribution of each cluster to the average dielectric constant (per site) is $\epsilon-1\propto1/q^2$, evaluated at $q\sim1/r_s$. This does not necessarily reproduce the result one should obtain by assuming zero field along the strands (i.e., electrostatic equilibrium). It is a low-field limit ($eEr_s/k_BT\ll1$).[21] This also applies to Refs. 3 and 4.

In general we expect the effect discussed here to be dominant when ϵ_2 is very small or when motion *within* clusters is dominant; e.g., at high frequencies. A crossover between the two behaviors is thus predicted as a function of either ω or ϵ_2. This crossover should show up through an anomalous behavior in high-frequency optical and electrical measurements.[22]

Finally, we note that similar anomalous phenomena were recently predicted to occur as a result of quantum localization.[14] When diffusion was studied within a single electron wave packet, with localization length ξ_l, it was found that $\epsilon\propto\xi_l^2$. However, the exponent θ was found to be $\theta_{loc}=d-2$, with values which are quite distinct from ours. The crossover between quantum diffusion and the classical one discussed here remains for future investigation.[23]

We gratefully acknowledge stimulating discussions with D. J. Bergman, M. H. Cohen, Y. Imry, S. Kirkpatrick, R. Landauer, M. J. Stephen, and J. P. Straley. One of us (A.A.) acknowledges the hospitality given him at IBM and the comments of E. Pytte. This work was supported by a grant from the Israel Academy of Arts and Sciences.

[1]For a detailed review see, e.g., D. Stauffer, Phys. Rep. 54, 3 (1979).

[2]S. Kirkpatrick, in *Ill-Condensed Matter,* Proceedings of the Les Houches Summer School, Session XXXI, edited by R. Balian, R. Maynard, and G. Toulouse (North-Holland, Amsterdam, 1979), and in *Electrical Transport and Optical Properties of Inhomogeneous Media —1977,* edited by J. C. Garland and D. B. Tanner, AIP Conference Proceedings No. 40 (American Institute of Physics, New York, 1978), p. 99.

[3]A. L. Efros and B. I. Shklovskii, Phys. Status Solidi (b) 76, 475 (1976).

[4]D. J. Bergman and Y. Imry, Phys. Rev. Lett. 39, 1222 (1977).

[5]A. Kapitulnik and G. Deutcher, Phys. Rev. Lett. 49, 1444 (1982).

[6]B. B. Mandelbrot, *The Fractal Geometry of Nature* (Freeman, San Francisco, 1982), and *Fractals: Form, Chance and Dimension* (Freeman, San Francisco, 1977).

[7]Y. Gefen, A. Aharony, B. B. Mandelbrot, and S. Kirkpatrick, Phys. Rev. Lett. 47, 1771 (1981).

[8]An alternative approach would be to consider explicitly the scaling properties of the relevant equations on the fractal geometry. This gives equivalent results (S. Alexander, to be published).

VOLUME 50, NUMBER 1 PHYSICAL REVIEW LETTERS 3 JANUARY 1983

[9]P. G. de Gennes, Recherche 7, 919 (1976).

[10]C. D. Mitescu, H. Ottavi, and J. Roussenq, in *Electrical Transport and Optical Properties of Inhomogeneous Media—1977*, edited by J. C. Garland and D. B. Tanner, AIP Conference Proceedings No. 40 (American Institute of Physics, New York, 1978), p. 377; J. Roussenq, thesis, Université de Provence, 1980 (unpublished).

[11]T. Vicsek, Z. Phys. B 45, 153 (1981).

[12]J. P. Straley, J. Phys. C 13, 2991 (1980).

[13]B. D. Hughes, M. F. Shlesinger, and E. W. Montroll, Proc. Natl. Acad. Sci. USA 78, 3287 (1981).

[14]A similar dependence on scale of a diffusion coefficient was recently stated in the context of the electron localization problem by Y. Imry, J. Appl. Phys. 52, 1817 (1981), and was employed to calculate the dielectric properties by Y. Imry, Y. Gefen, and D. J. Bergman, Phys. Rev. B 26, 3436 (1982), and unpublished.

[15]It is interesting to note that for $p < p_c$ and $r_s \gg \xi$ one expects the large finite clusters to scale like "animals" (Ref. 1). The form (2) may then yield new diffusion laws. A new exponent θ_a may then replace θ in Eq. (1) when $\xi \ll r(t) \ll r_s$. In the present paper we ignore these (exponentially rare) cases.

[16]Following the present work, the authors of Ref. 10 looked again at their short-time data and found $\langle r^2(t)\rangle \sim t^{0.66}$ ($d=3$) and $\langle r^2(t)\rangle \sim t^{0.8}$ ($d=2$). Considering the error bars, and the fact that for very short times $\langle r^2(t)\rangle \propto t^2$, biasing the measured exponents to higher effective values, these preliminary results are probably not inconsistent with our predictions.

[17]M. J. Stephen, Phys. Rev. B 17, 4444 (1978).

[18]J. P. Straley, Phys. Rev. B 15, 5733 (1977).

[19]J. P. Straley, in *Electrical Transport and Optical Properties of Inhomogeneous Media—1977*, edited by J. C. Garland and D. B. Tanner, AIP Conference Proceedings No. 40 (American Institute of Physics, New York, 1978), p. 118; C. D. Mitescu, M. Allain, E. Guyon, and S. P. Clerc, J. Phys. A 15, 2523 (1982).

[20]I. Webman, Phys. Rev. Lett. 47, 1496 (1981).

[21]Note that the result $P \propto r_s^2 E$ follows directly if the change in the charge density, $\Delta\rho$, is proportional to the electric potential Er_s. This is to be expected, e.g., at low field, when $eEr_s/k_BT \ll 1$, when $\Delta\rho \simeq 1 - eEr_s/k_BT$. In principle it could also apply at higher fields and sufficiently high frequencies, but the conditions for this are rather stringent and complex.

[22]For $d=3$ gold samples, characterized by grains of size $a \sim 100$ Å, with $\xi \sim 50a$, at $T \sim 300$ K, we estimate $D(\xi) \sim 10^{-2}$ cm^2/sec. The characteristic crossover frequency is thus $\tilde{\omega} = (A/\xi)^{2+\theta} \sim 10^7$ sec^{-1}. The trivial upper limit is $1/\tau_0 \sim 10^{14}$ sec^{-1}, leaving a wide frequency range for experiments. Details will be published elsewhere.

[23]Quantum localization behavior is expected for $\omega \gtrsim 10^{12}$ sec^{-1}; see Ref. 14.